Advances in Industrial Control

Springer
London
Berlin
Heidelberg
New York
Barcelona
Budapest
Hong Kong
Milan
Paris
Santa Clara
Singapore
Tokyo

Other titles published in this Series:

Parallel Processing for Jet Engine Control
Haydn A. Thompson

Iterative Learning Control for Deterministic Systems
Kevin L. Moore

Parallel Processing in Digital Control
D. Fabian, Garcia Nocetti and Peter J Fleming

Intelligent Seam Tracking for Robotic Welding
Nitin Nayak and Asok Ray

Identification of Multivariable Industrial Process for Simulation, Diagnosis and Control
Yucai Zhu and Ton Backx

Nonlinear Process Control: Applications of Generic Model Control
Edited by Peter L. Lee

Microcomputer-Based Adaptive Control Applied to Thyristor-Driven D-C Motors
Ulrich Keuchel and Richard M. Stephan

Expert Aided Control System Design
Colin Tebbutt

Modeling and Advanced Control for Process Industries, Applications to Paper Making Processes
Ming Rao, Qijun Xia and Yiquan Ying

Robust Multivariable Flight Control
Richard J. Adams, James M. Buffington, Andrew G. Sparks and Siva S. Banda

Modelling and Simulation of Power Generation Plants
A.W. Ordys, A.W. Pike, M.A. Johnson, R.M. Katebi and M.J. Grimble

Model Predictive Control in the Process Industry
E.F. Camacho and C. Bordons

H_∞ Aerospace Control Design: A VSTOL Flight Application
R.A. Hyde

Advances in Neural Networks for Control Systems
Edited by Kenneth Hunt, George Irwin and Kevin Warwick

Sigeru Omatu, Marzuki Khalid and Rubiyah Yusof

Neuro-Control and its Applications

With 88 Figures

Springer

Sigeru Omatu, PhD, Professor

Department of Computer and Systems Sciences, College of Engineering, Osaka Prefecture University, Sakai, Osaka 593, Japan

Marzuki Khalid, PhD, Associate Professor
Rubiyah Yusof, PhD, Lecturer

Business and Advanced Technology Centre, Universiti Teknologi Malaysia, Jalan Semarak, 54100 Kuala Lumpur, Malaysia

Series Editors

Michael J. Grimble, Professor of Industrial Systems and Director
Michael A. Johnson, Reader in Control Systems and Deputy Director

Industrial Control Centre, Department of Electronic and Electrical Engineering, Graham Hills Building, 60 George Street, Glasgow G1 1QE, UK

ISBN-13:978-1-4471-3060-4 e-ISBN-13:978-1-4471-3058-1
DOI: 10.1007/978-1-4471-3058-1

British Library Cataloguing in Publication Data
Neuro-Control and Its Applications. -
(Advances in Industrial Control Series)
I. Title II. Series
629.89
ISBN-13:978-1-4471-3060-4

Library of Congress Cataloging-in-Publication Data
Omatu, S., 1946-
Neuro-control and its applications / Sigeru Omatu, Marzuki Khalid, and Rubiyah Yusof.
p. cm. -- (Advances in industrial control)
Includes bibliographical references and index.
ISBN-13:978-1-4471-3060-4 (alk. paper)
1. Intelligent control systems. 2. Neural networks (Computer science) I. Marzuki Khalid, 1959- . II. Rubiyah Yusof, 1961-
III. Title. IV. Series.
TJ217.5.053 1995 95-20218
670.42--dc20 CIP

Softcover reprint of the hardcover 1st edition 1996

Typesetting: Camera ready by authors

69/3830-543210 Printed on acid-free paper

SERIES EDITORS' FOREWORD

The series *Advances in Industrial Control* aims to report and encourage technology transfer in control engineering. The rapid development of control technology impacts all areas of the control discipline. New theory, new controllers, actuators, sensors, new industrial processes, computer methods, new applications, new philosophies, ..., new challenges. Much of this development work resides in industrial reports, feasibility study papers and the reports of advance collaborative projects. The series offers an opportunity for researchers to present an extended exposition of such new work in all aspects of industrial control for wider and rapid dissemination.

Sigeru Omatu, Marzuki Khalid, and Rubiyah Yusof have pursued the new developments of fuzzy logic and neural networks to present a series volume on neuro-control methods. As they demonstrate in the opening pages of their book, there is an explosion of interest in this field. Publication and patent activity in these areas are ever growing according to international databases and hence, this volume is timely.

The presentation of the material follows a complementary pattern. Reviews of existing control techniques are given along side an exposition of the theoretical constructions of fuzzy logic controllers, and controllers based on neural networks. This is an extremely useful methodology which yields rewards in the applications chapters. The series of applications includes one very thorough experimental sequence for the control of a hot-water bath. Here, the results from four controllers (including PI and GPC) are compared. The results from this and the other applications are quite thought provoking, rounding out a useful and comprehensive volume.

M.J. Grimble and M.A. Johnson
Industrial Control Centre
Glasgow, Scotland

PREFACE

Recent progress in neural network theory and fuzzy logic theory provides us with new tools for modeling, estimation, and control of complex nonlinear systems while traditional and conventional control theory does the same for only linear systems. In order to achieve results that not only control theorists but also industrial practitioners can use to solve difficult real control problems which are basically nonlinear, noisy, and complex, this book provides hands-on approaches to intelligent control, with special focus on neuro-control.

Neuro-control is a control scheme using artificial neural networks which are based on the biological brain information processing mechanism. The neural network approach has superior characteristics when compared with conventional computers. Some of them are; (1) easier pattern information processing, (2) self-organization, (3) distributed memory, (4) parallel processing architecture, and (5) learning. However, conventional computers of the Von Neumann type have the following properties; (1') symbolic expression, (2') logical representation, (3') local memory, (4') serial processing architecture, and (5') sequential algorithmic programming. Therefore, if we use neural networks for control, we can expect that the neuro-controllers have at least some of the properties and advantages as discussed above. To this end, research works on neuro-control have already achieved some of the objectives and features of intelligent control such as parallel and distributed processing, and self-organizing and learning capabilities.

This book discusses various types of neuro-control paradigms based on the backpropagation algorithm which is a supervised learning technique derived by the steepest descent (gradient) method. The neuro-controller can realize a nonlinear control algorithm which is robust to noise, complexities, and variations in the plant. Some of the examples on the applications of the neuro-control techniques to real control problems given in this book should enable the reader to have some idea of how the techniques are implemented. The neuro-control methodologies and their experimental performances are also compared with those of conventional and traditional control methods. This would enable the reader to understand the similarities and differences between the new neuro-control approach and the other control approaches. It would also allow the reader to decipher the advantages and disadvantages of the different control techniques and to select their best features and characteristics for the development of a truly intelligent and efficient control system. In due course, these new neuro-control techniques will be available to domestic appliances and industrial control systems, similar to the fuzzy logic revolution in recent years.

This book has been organized as follows: Chapter 1 presents an overview of intelligent control and more specifically introduction to the definition of neuro-control. Chapter 2 introduces the reader some aspects of biological neural networks and their relation to artificial neural networks. The backpropagation algorithm is thoroughly discussed in this chapter which forms the main algorithm of the neuro-control schemes presented in this book.

Chapter 3 deals with traditional control techniques which include several adaptive control techniques and also fuzzy logic control. This chapter is intended to allow the reader to understand some basic concepts of the more popular control techniques that are available today such as self-tuning control, generalized predictive control, and fuzzy logic. The discussion on these traditional control techniques can be used as a basis of comparison to the neuro-control techniques that are discussed in Chapter 4. Chapter 4 starts with an overview of the neuro-control techniques that have been much cited in the literature. The rest of the chapter discusses various neuro-control schemes from our viewpoint based on the backpropagation algorithm. Several simulation examples based on some of the neuro-control paradigms are given in this chapter. Chapter 5 deals specifically with four different neuro-control applications. Developments of the neuro-control schemes on real processes are described. The performances of these neuro-controllers are also compared with several traditional control schemes implemented on the same processes. From the experimental results the neuro-control schemes show considerable robustness and encouraging advantages not found in many traditional control approaches. We believe that these neuro-control schemes have strong potential for many real world applications in the near future.

We have also included a program list written in Borland C++ (runs on Borland C++ Version 3.1 Environment). The program can be used to train a multilayered neural network model the inverse dynamics of a plant from its input-output characteristics. After training the neural network can be configured to be used as a direct controller to the plant.

ACKNOWLEDGEMENTS

We would like to thank Professor M. J. Grimble and Dr. M. A. Johnson at the University of Strathclyde who have recommended us to write the book. Appreciation goes to our former students: Dr. Shigenori Matsumura who developed the self-tuning PID neuro-controller application to an electric vehicle, Mr. Saiful Akhyar who considered the basic structure of self-tuning PID neuro-controller, Mr. Julio Tanomaru who contributed the inverse dynamic controller design, and Mr. Tatsuya Ide who realized the fuzzy neuro-controller for the stabilization of an inverted pendulum.

We would also like to convey our appreciation to the Japan Society for the Promotion of Science (JSPS) and the Vice Chancellor's Council of Malaysian Universities (VCC) for financing several of our trips between Japan and Malaysia which contribute to the possible completion of the book.

Finally, we wish to thank Mr. Nicholas Pinfield and Mrs. Imke Mowbray who serve as Engineering Editor and Editorial Assistant for Springer-Verlag, respectively, and have given fruitful suggestions towards the completion of the book at Kuala Lumpur, Malaysia, 1995.

Sigeru Omatu

University of Tokushima, Japan.

Marzuki Khalid
Rubiyah Yusof

University of Technology, Malaysia.

CONTENTS

4 Neuro-Control Techniques

5 Neuro-Control Applications

CHAPTER 1
INTRODUCTION

1.1 Introduction to Intelligent Control

Intelligent control is now becoming a common tool in many engineering and industrial applications [1], [2]. It has the ability to comprehend and learn about plants, disturbances, environment, and operating conditions [3],[4]. Some examples of the factors to be learned are plant characteristics such as its static and dynamic behaviours, some characteristics of disturbances or the environment, and equipment-operating practices [1], [2]. Figures 1.1.1 and 1.1.2 show the number of papers from INSPEC (Information Service for Physics and Engineering Communities) and patents from CASSIS (Classification for Search Support Information System), respectively [5]. From these figures, we can see the trends and the relative activities of research and applications in the field of computational and artificial intelligence. It can be observed that while research in expert systems which used to be the domain tool for intelligent systems, is declining slowly, research in neural networks is progressing rather steadily.

Artificial neural networks, with their self-organizing and learning ability, are now used as promising tools for such purposes [3], [4]. The architecture and functions of the artificial neural networks are based on the biological brain. Neural networks provide a different computing architecture compared with the Von Neumann computers. The main characteristics of neural networks are parallel and distributed in nature as well as self-organization, however, conventional computers have series, local, and algorithmic properties [3], [4]. Parallel architecture means that information processing is done by many central processing units (CPUs), while in a conventional computer which has only one CPU, the information is processed in a sequential step by step manner.

The use of distributed memory means that information is stored into many addresses in a distributed way such that each entity is represented by a pattern of activity distributed over many computing elements and each computing element is involved in representing many different entities. Conventional computers use local memory or local representation which uses one computing element for each entity. Adopting the distributed representation architecture in neural networks, information can be broken up and processed into many different parts [3].

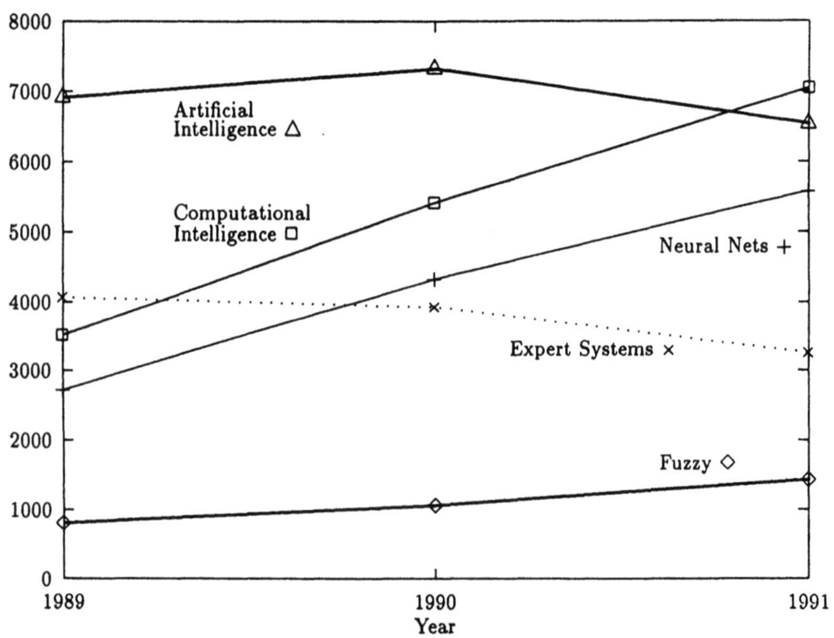

Fig. 1.1.1. Numbers of papers and other works generated in areas of computational and artificial intelligence according to the INSPEC database by Zurada *et al.*[5] (© 1994 by IEEE Press)

The last property of self-organization is called learning [1]. This means that neural networks can learn static or dynamic properties autonomously based on the past history of measurement data and then act in a way such that a better solution can be obtained under unknown environmental conditions. However, conventional computers have to be programmed before data can be processed and they cannot work beyond the decision given by the program. Therefore, knowledge-based engineering has not been well accepted in real applications since it has no solution when it has to make a decision under new environments [1].

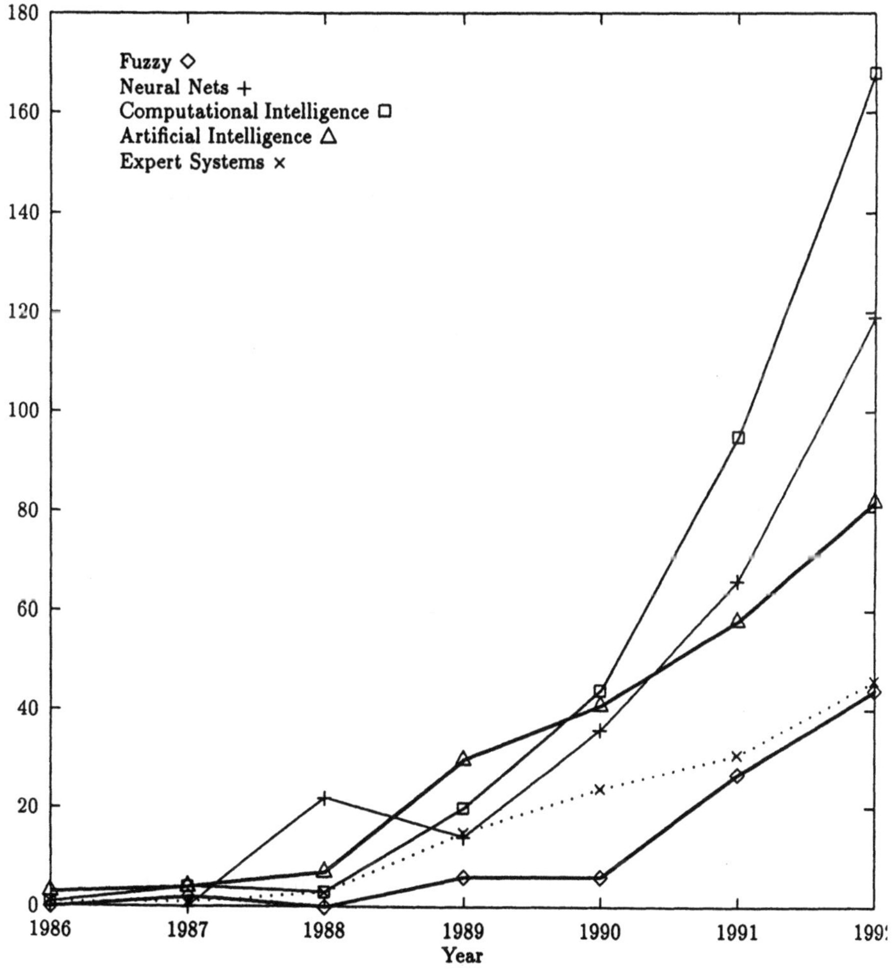

Fig. 1.1.2. Patents activity according to the CASSIS data base by Zurada *et al.*[5] (© 1994 by IEEE Press).

Neuro-control schemes are those which use neural network architecture's and learning capabilities. Neural networks consist of neuron-like computing elements which are basically nonlinear. These nonlinear properties of neural networks allow the possibility of nonlinear mappings and thus, neuro-control can realize new nonlinear control schemes. In reality, various kinds of nonlinear neuro-control schemes can be developed to solve some of our unsolved control problems. One of the reasons why a lot of attentions have been paid to neuro-control is that conventional and traditional control methodologies are mainly based on linear systems theory while real plants are in effect nonlinear in nature and have unmodeled dynamics. Plant designers have developed systems not from a control theoretical viewpoint but rather from the feasibility of the design. Thus, control engineers must decide the control policy to achieve the most efficient productivity under many constraints of a real plant.

In real operation, it is often difficult to make the plant fully automatic and we need operators to control it. Even if we design a highly sophisticated control scheme without the plant operator being able to understand its technique and principles, its reliability or safety is low since there may be unpredictable phenomena occurring in real operation. This problem can lead to undesirable loss in raw materials as well as the possibility of accidents occurring in the plant. Due to this problem, the popular proportional-plus-integral-plus-derivative (PID) controller has been widely accepted. Some reasons for its wide popularity are its simple principle structure and high reliability. According to a recent survey [6], eighty-four per cent of Japanese industry is still using the conventional PID controller.

In these types of controllers, three parameters, which are the PID gains, are needed to be varied and set by the operators. Due to its simplicity and reliability, the PID controller is very much acceptable by the industrial control community. However, there are a number of drawbacks in this conventional controller such that when the operating points have been changed they need to be re-tuned again. In large plants, where they have to be operated continuously, close monitoring is essential when using this controller which needs significant human resources.

In this light, new control schemes should be devised for a better solution to our control needs. Moreover, new control schemes should be simple enough in principle and operation in order to be readily accepted by the industrial control community at large. Perhaps, it may be difficult to design the control of large scale systems using only simple control principles in order to attain a high level of performance. However, it is very much desirable that minimum manipulation of parameters to be an inherent feature in the new control schemes. Thus, a simple but effective man-machine interface is an important factor for the development of intelligent control systems.

In the context of engineering, intelligent control must have the following features; (a) learning ability and adaptability, (b) robustness, (c) simple control algorithm for relatively user-friendly man-machine interface, and (d) not exclusive but inclusive nature to realize better solution under constraints of hardware equipments. Currently, there are a number of techniques that can be used as a basis for the development of intelligent systems, namely, expert or knowledge-based systems, fuzzy logic, genetic algorithms, neural networks, and artificial life, to name the more popular ones. These artificial intelligent (AI) techniques should be integrated with modern control theory to develop intelligent control systems.

In this book we introduce the concept of neural networks and their applications to control systems. The developments of the neuro-control systems in this book are based entirely on the backpropagation algorithm. This algorithm has been used widely and successfully in many engineering applications. Hence, it has been our choice of learning algorithm for the neural networks in our applications. First, the concept of artificial neural networks is presented which is then followed by the various kinds of neuro-control paradigms. In order to see how well the neuro-control systems perform we make comparisons with conventional and traditional control approaches. These conventional and traditional control approaches are discussed in the following chapter which include PID controllers, self-tuning controllers, generalized predictive controllers, and also fuzzy logic controllers. Their concepts and approaches can then be readily compared to the neuro control schemes which are applied on the same control systems which are discussed in the final chapter.

1.2 References

[1] White, D.A. and D.A Sofge, *"Handbook of Intelligent Control"*, Van Nostrand Reinhold, New York, 1992.

[2] Miller, W.T., R.S. Sutton, and P.J. Werbos, *"Neural Networks for Control"*, MIT Press, Cambridge, MA, 1990.

[3] Rumelhart, D.E. *et al.*, *"Parallel Distributed Processing: Explorations in the Micro Sructure of Cognition Vol.I: Foundations"*, MIT Press, Cambridge, MA, 1986.

[4] Dayhoff, J.E., *"Neural Network Architectures: An Introduction"*, Van Nostrand Reinhold, New York, 1990.

[5] Zurada, J.M., R.J. Marks II, and C.J. Robinson, *"Computational Intelligence Imitating Life"*, IEEE Press, New York, 1994.

[6] Nozaka, Y., "Trend of new control theory application in industrial process control (A survey)", *Proc. of 12th IFAC World Congress*, Sydney, Vol. VI, pp. 51-56, 1993.

CHAPTER 2

NEURAL NETWORKS

2.1 Historical Review of Neural Networks

Neural networks are networks of nerve cells (neurons) in the brain. The human brain has billions of individual neurons and trillions of interconnections. Neurons are continuously processing and transmitting information to one another. In 1909, Cajal [1], [2] found that the brain consists of a large number of highly connected neurons which apparently can send very simple excitatory and inhibitory messages to each other and update their excitations on the basis of these simple messages. Figure 2.1.1 shows Purkinje Cell with its dendrite stained [2]. A neuron has three major regions; the cell body (soma), the axon, and the dendrites as shown in Fig. 2.1.2 [2].

The soma provides the support functions and structure of the cell. The axon is a branching fiber which carries signals away from the neurons. The dendrites consist of more branching fibers which receive signals from other nerve cells. The outgoing signals are pulses which are initiated at the axon hillock where the axon connects to the soma. Figure 2.1.3 [2] shows a schematic diagram of a synapse, which includes the pre-synaptic cell, its synaptic vesicles, the synaptic cleft, and the post-synaptic cell.

A terminal bouton is at the end of the axon branch as shown in Fig 2.1.2. The structure releases chemical transmitters which cross the synaptic gap to reach a receiving neuron. When the transmitters are released, the membrane of the target neuron is affected and its inclination to fire its own impulse is either increased or decreased according to the condition whether the incoming signal is either excitatory or inhibitory. Since the synapse is a keystone of artificial neural network, we will explain about it in more detail. Figure 2.1.3 shows a schematic diagram of a synapse where the pre-synaptic cell, its synaptic vesicles, the synaptic cleft, and the post-synaptic cell are included. In the pre-synaptic cell, incoming signals affect the membrane in the pre-synaptic area, and cause a release of a transmitter from the synaptic vesicles. The transmitter crosses the synaptic cleft to reach the post-synaptic cell, which causes the membrane properties of the receiving cell. The firing rate of the post-synaptic cell is either increased or decreased.

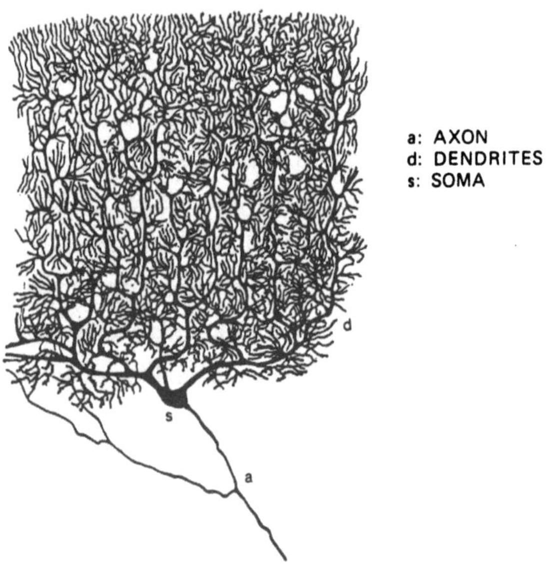

Fig. 2.1.1. Purkinje Cell (from the cerebellum) with its dendrites stained; they form a flat fan adopted from J.E. Dayhoff [2] (© re-printed with permission from Int. Thomson Publishers).

The pre-synaptic cell has a button at the end of its axonal process. The arrival of the action potential affects the pre-synaptic cell membrane at the button terminal. The membrane responds by allowing an influx of calcium ions (Ca^{++}), then the transmitter is released from the synaptic vesicles into the cleft. The release of the transmitter appears to be modulated by Ca^{++}. Ca^{++} ions are involved in candidate mechanisms for the modulation of synaptic strength during learning. It is not the intention of this book to delve deeply into synaptic vesicles, transmitter substances, post-synaptic cell, etc. A detailed description of this area can be found in the book by J.E. Dayhoff [2].

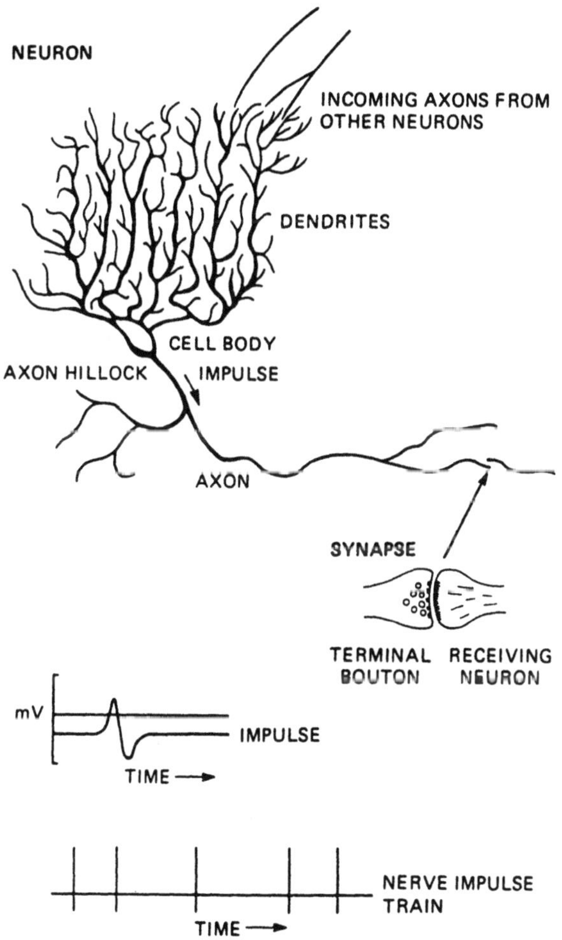

Fig. 2.1.2. Recording a nerve impulse train with a micro electrode placed near an axon. A trace of the nerve wave form is shown, as it would appear on an oscilloscope if recorded from the micro electrode adopted from J.E. Dayhoff [2] (© re-printed with permission from Int. Thomson Publishers).

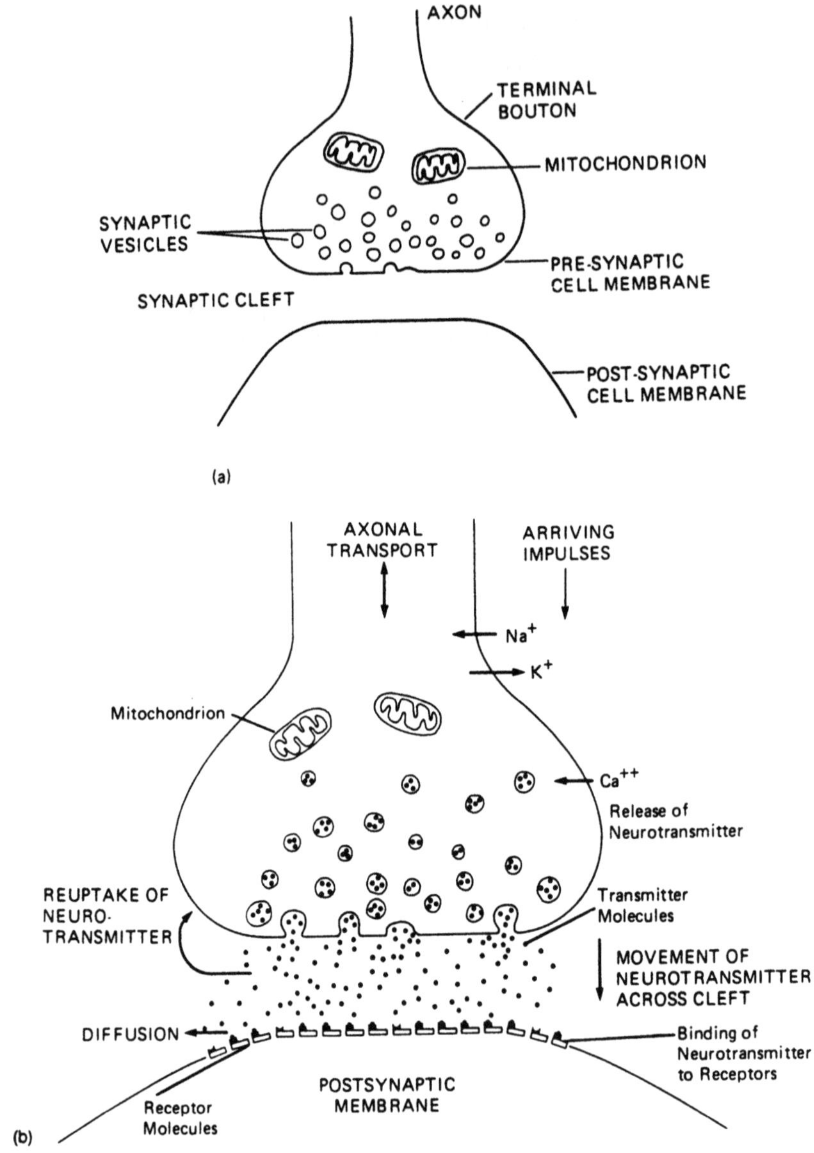

Fig. 2.1.3. (a) Schematic diagram of anatomical structures in the synapse. (b) Dynamic processes of the synapse adopted from J.E. Dayhoff [2] (© re-printed with permission from Int. Thomson Publishers).

In 1943, McCulloch and Pitts proposed a mathematical model of the neurons and showed how neuronal-like networks could be computed. Figure 2.1.4 shows a schematic model of the McCulloch-Pitts neuron and its activity.

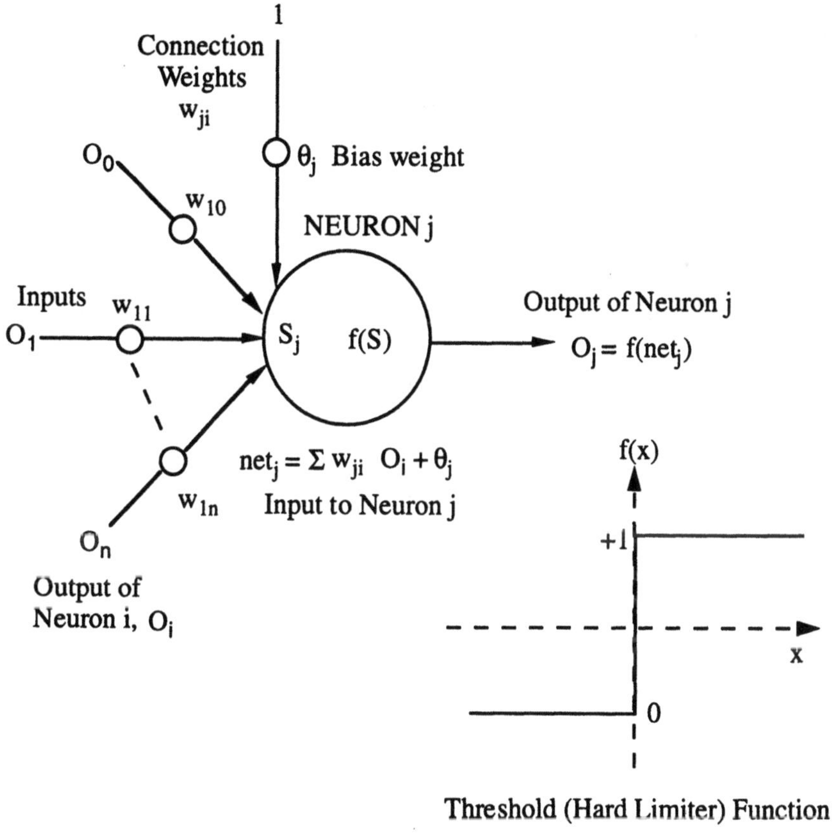

Fig. 2.1.4. Schematic model of a McCulloch-Pitts neuron and its activity.

The mathematical model of the neuron by McCulloch and Pitts is described as

$$net_j = \sum_i w_{ji} O_i + \theta_j \tag{2.1.1}$$

$$O_j = f(net_j) \tag{2.1.2}$$

where net_j is the net input to a neuron j, θ_j is a threshold value, w_{ji} is the strength and sense of synaptic connection from a neuron i to a neuron j, O_i is an output signal of a neuron i, and f(x) is an output function or an activation function of a neuron. For simplicity, we use the following types of f(x):

(i) hard limiter or sign function:

$$f(x) = \text{sign}\,(x) = \begin{cases} 1 & x \geq 0 \\ 0 & x < 0 \end{cases}$$

(ii) linear function : $f(x) = x$

(iii) sigmoid function : $f(x) = \text{sigmoid}\,(x) = \dfrac{1}{1 + \exp\left(-\dfrac{x}{T}\right)}$, $T > 0$.

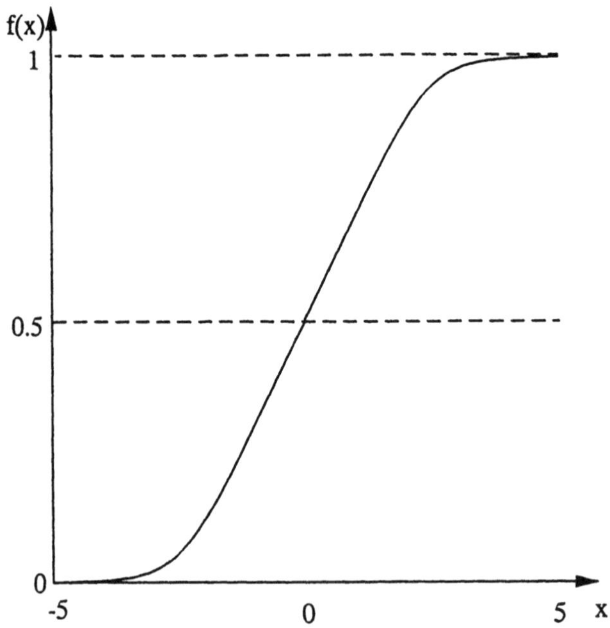

Fig. 2.1.5. A sigmoid logistic activation function.

For the sigmoid function, note that if $T \to \infty$, then f(x) goes to the linear function and if $T \to 0$, then f(x) goes to the sign function as shown in Fig. 2.1.5.

The first set of ideas of learning in neural networks was contained in Hebb's book [4] entitled *The Organization of Behaviour* in 1949. Before Hebb's work, it was believed that some physical change must occur in a neural network to support learning, however, it could not be determined what this change was. Hebb assumed that a reasonable biological change would be needed to strengthen the connections between elements of the neural network only when both the pre-synaptic and post-synaptic cell membranes were active simultaneously.

The essence of Hebb's ideas occurs in various learning paradigms. Although the details of the rules for changing the weights may be different, Hebb's essential notion that the strength of connections between the units must change in response to some function of the correlated activity of the connected units has been adopted in many learning models.

In 1951, Edmonds and Minsky built their learning machine using Hebb's idea [5]. Although Minsky was the first person to propose a learning machine, the real beginning of a meaningful neuron-like network learning can be traced to the work of Rosenblatt in 1962 [6]. Rosenblatt invented a class of simple neuron-like learning networks which is called perceptron. The perceptron was actually an entire class of architecture which was composed of processing units that transmitted signals and adapted their connection weights. Rosenblatt's research was oriented toward modeling the brain in an attempt to understand memory, learning, and cognitive processes. Rosenblatt's works were extended by many scientists and engineers and a number of machines were built based on perceptron architectures. They worked on a mathematical analysis of the computing powers of perceptrons with the goal of understanding these limitations. Minsky and Papert's book [7] had a damping effect on the study of neuron-like networks as computational devices.

In a breakthrough paper published in 1982, Hopfield introduced a neural network architecture which is called Hopfield network [8]. He described how computational capabilities can be built from neural networks. He illustrated an associative memory that can be implemented with this network, and later demonstrated that the network can be used to solve optimization problems such as the traveling salesman problem [9].

Another approach which has been widely used in the neural network paradigms is the back-error-propagation or backpropagation. The conceptual basis of backpropagation was first presented in 1974 by Werbos [10], then was independently re-invented in 1986 by Rumelhart *et al.* [11]. Their book, *Parallel Distributed Processing*, introduced a broad perspective of the neural network approaches.

Other recent developments are Boltzmann machine, competitive learning, and Kohonen's feature map. The Boltzmann machine was developed by Hinton and Sejnowski based on thermodynamics models [11]. The thermodynamics models employ binary units which take on the values $\{0,1\}$. They are determined by a stochastic function of the inputs. Under similar assumptions to the Hopfield network, it can be proved that the network moves into those states that are maximally consistent with the input and with the internal constraints represented by the weights.

The competitive learning paradigm uses a layer of processing units that compete with one another, resulting in a neural network that can be used for pattern-classification application, which was considered by Rumelhart and Zipser [12] and Grossberg [13]. Kohonen's self-organizing feature map is a two-layered network that can organize a topological map from a random starting point. The resulting map shows the natural relationships among the patterns that are given to the network. This paradigm was proposed by Kohonen [14]. The results presented in this book are mainly based on the backpropagation algorithm. This backpropagation algorithm will be explained in the next section.

2.2 Backpropagation Algorithm

The backpropagation algorithm is an extension of perceptrons or ADALINE [15] to multilayered neural networks. Thus, the backpropagation algorithm employs three or more layers of processing units (neurons). Figure 2.2.1 shows an architecture of a typical three-layered network for the backpropagation algorithm. The leftmost layer of units is the input layer to which the input data is supplied. The layer after it is the hidden layer where the processing units are interconnected to the layers before and after it. The rightmost layer is the output layer. The layers in Fig. 2.2.1 are fully interconnected, which means that each processing unit is connected to every unit in the previous layer and in the succeeding layer. However, units are not connected to other units in the same layer. Note that the backpropagation networks do not have to be fully interconnected and that any number of hidden layer may be used.

2.2.1 Notation

For simplicity, we consider three-layered neural networks given by Fig. 2.2.1, where O_k, O_j, and O_i are output values at the output, hidden, and input layers, respectively. A connection weight from a unit j at the hidden layer to a unit k at the output layer is denoted by w_{kj} . Similarly, w_{ji} is a connection weight from a unit i at the input layer to a unit j at the hidden layer. Thus, we have at the output layer

$$O_k = f(net_k) \tag{2.2.1}$$

$$net_k = \sum_j w_{kj} O_j + \theta_k \tag{2.2.2}$$

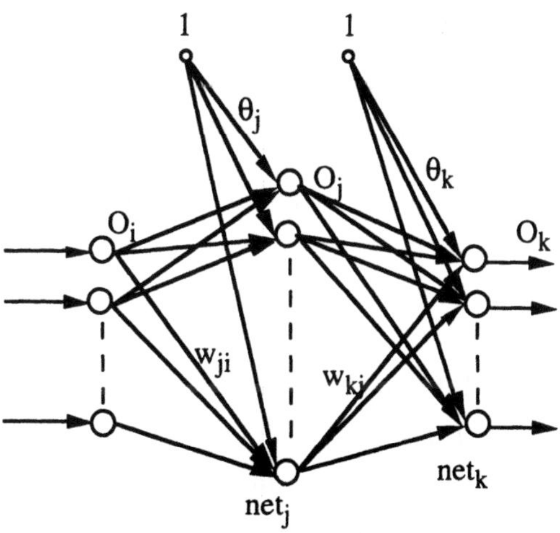

Fig. 2.2.1. A multilayered Perceptron

and at the hidden layer

$$O_j = f(net_j) \tag{2.2.3}$$

$$net_j = \sum_i w_{ji} O_i + \theta_j \tag{2.2.4}$$

where f(x) is given by

$$f(x) = \frac{1}{1 + \exp(-x)}. \tag{2.2.5}$$

The learning procedure involves the presentation of a set of pairs of input and output patterns. The neural network first uses the input pattern to produce its own output pattern and then compares this with the desired output or the target pattern. If there is no difference between an actual output and the target pattern, then no learning takes place. Otherwise, the connection weights are changed to reduce the difference. The criteria on function used here is a squared error function E given by

$$E = \sum_p E_p \tag{2.2.6}$$

$$E_p = \frac{1}{2} \sum_k (\tau_{pk} - O_{pk})^2 \tag{2.2.7}$$

where τ_{pk} is the desired output for the *kth* component of the output pattern for pattern p and O_{pk} is the corresponding actual output. Thus, E_p is a measure of the error on input/output for pattern p and E is the overall measure of the error.

2.2.2 Derivation of the Backpropagation Algorithm

Based on the steepest descent (gradient) method, the derivation of the backpropagation algorithm is shown in this section. This derivation shows how the connection weights between the layers of the multilayered neural network are adapted. We begin by first determining the error function and the criteria for the adaptation of the weight which is proportional to the ratio of the derivative of the error measure with respect to each weight. First, we derive the backpropagation algorithm for a specific pattern p, that is, find a minimizing sequence of E_p. For simplicity, we omit the subscript p for a while and hence,

$$E_p = \frac{1}{2} \sum_k (\tau_k - O_k)^2. \tag{2.2.8}$$

2.2.2.1 Output Layer Rule

According to the steepest descent (gradient) method,

$$\Delta w_{kj} \propto -\frac{\partial E_p}{\partial w_{kj}}$$

where Δw_{kj} is the change to be made to the connection weight w_{kj}, that is,

$$\Delta w_{kj} = w_{kj}(\text{new}) - w_{kj}(\text{old}). \tag{2.2.9}$$

This adaptation can be described in Fig. 2.2.2. Note that in Fig. 2.2.3 it can be observed that the steepest descent (gradient) method may sometimes approach a local minimum point. Therefore, we get

$$\Delta w_{kj} = -\eta \frac{\partial E_p}{\partial w_{kj}}, \quad \eta > 0 \tag{2.2.10}$$

where η is a learning rate.

Let us introduce the generalized error signal δ_k given by

$$\delta_k = -\frac{\partial E_p}{\partial net_k}. \tag{2.2.11}$$

Then using the chain rule yields

$$\frac{\partial E_p}{\partial w_{kj}} = \frac{\partial E_p}{\partial net_k} \frac{\partial net_k}{\partial w_{kj}} = -\delta_k \frac{\partial net_k}{\partial w_{kj}}. \tag{2.2.12}$$

This means that $\frac{\partial E_p}{\partial w_{kj}}$ results from the product of two parts, one part reflecting the change in error as a function of the change in the net input to the unit and one part representing the effect of changing a particular connection weight on the net input. From (2.2.2) we have

$$\frac{\partial net_k}{\partial w_{kj}} = \frac{\partial\left(\sum_j w_{kj} O_j + \theta_k\right)}{\partial w_{kj}} = O_j$$

and hence,

$$\Delta w_{kj} = \eta \delta_k O_j. \tag{2.2.13}$$

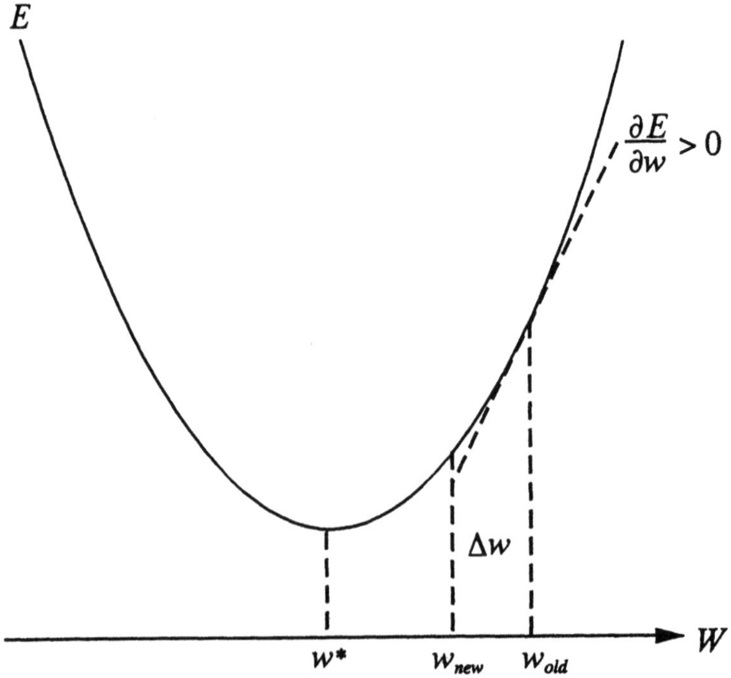

Fig. 2.2.2. Principle of the steepest descent (gradient) method.

To compute δ_k, we apply the chain rule. Thus, we get

$$\delta_k = -\frac{\partial E_p}{\partial net_k} = -\frac{\partial E_p}{\partial O_k}\frac{\partial O_k}{\partial net_k}. \tag{2.2.14}$$

This means that δ_k is the product of two factors; one factor reflecting the change in error with respect to the output of the unit and one factor representing the change in the output with respect to changes in the net input. From (2.2.8) we have

$$\frac{\partial E_p}{\partial O_k} = -(\tau_k - O_k) \tag{2.2.15}$$

and from (2.2.1)

$$\frac{\partial O_k}{\partial net_k} = f'(net_k) \tag{2.2.16}$$

where $f'(x)$ denotes the derivative of $f(x)$ with respect to x.
From (2.2.5) we have

$$f'(x) = \frac{e^{-x}}{(1+e^{-x})^2} = \frac{1}{(1+e^{-x})}\left(1 - \frac{1}{1+e^{-x}}\right)$$

$$= f(x)(1 - f(x)) \tag{2.2.17}$$

and hence,

$$\frac{\partial O_k}{\partial net_k} = O_k(1 - O_k). \tag{2.2.18}$$

Thus, we have

$$\Delta w_{kj} = \eta \delta_k O_j$$

$$\delta_k = O_k(1 - O_k)(\tau_k - O_k). \tag{2.2.19}$$

2.2.2.2 *Hidden Layer Rule*

As shown in the preceding section, the changing rule of the connection weights at the hidden layer is given by using the steepest descent (gradient) method as follows:

$$\Delta w_{ji} = -\eta \frac{\partial E_p}{\partial w_{ji}}. \tag{2.2.20}$$

Using the chain rule as before, we have

$$\frac{\partial E_p}{\partial w_{ji}} = \frac{\partial E_p}{\partial net_j} \frac{\partial net_j}{\partial w_{ji}}. \tag{2.2.21}$$

Let us define δ_j by

$$\delta_j = -\frac{\partial E_p}{\partial net_j}. \tag{2.2.22}$$

Since from (2.2.4), we get

$$\frac{\partial net_j}{\partial w_{ji}} = O_i$$

and hence,

$$\Delta w_{ji} = \eta \delta_j O_i. \tag{2.2.23}$$

Using the chain rule, we get

$$\delta_j = -\frac{\partial E_p}{\partial net_j} = -\sum_k \frac{\partial E_p}{\partial net_k} \frac{\partial net_k}{\partial O_j} \frac{\partial O_j}{\partial net_j}. \tag{2.2.24}$$

Note that if we change net_j, then all the outputs O_k are affected and hence, the summation with respect to k is required in the above chain rule. Using (2.2.11), (2.2.2), (2.2.3), and (2.2.5), we have

$$\delta_j = \sum_k \delta_k w_{kj} f'(net_j)$$

$$= O_j(1 - O_j) \sum_k \delta_k w_{kj} \,. \tag{2.2.25}$$

Therefore, we have obtained the following algorithm to change the connection weights.

$$\Delta w_{kj} = \eta \delta_k O_j \tag{2.2.26}$$

$$\Delta w_{ji} = \eta \delta_j O_i \tag{2.2.27}$$

$$\delta_k = O_k(1 - O_k)(\tau_k - O_k) \tag{2.2.28}$$

$$\delta_j = O_j(1 - O_j) \sum_k \delta_k w_{kj} \,. \tag{2.2.29}$$

In order to increase the learning rate without leading to oscillation in the output response, we modify the above equations (2.2.6) and (2.2.7) to include a momentum term. This can be accomplished by the following rule:

$$\Delta w_{kj}(t + 1) = \eta \delta_k O_j + \alpha \Delta w_{kj}(t) \tag{2.2.30}$$

$$\Delta w_{ji}(t + 1) = \eta \delta_j O_i + \alpha \Delta w_{ji}(t) \tag{2.2.31}$$

where t indexes the presentation time, η is the learning rate, and α is a constant which determines the effect of past connection weight changes on the current direction of movement in connection weights space. Now, combining these equations with the observation that

$$\frac{\partial E}{\partial w_{kj}} = \sum_p \frac{\partial E_p}{\partial w_{kj}} \text{ or } \frac{\partial E}{\partial w_{ji}} = \sum_p \frac{\partial E_p}{\partial w_{ji}} \tag{2.2.32}$$

should include that the net change in the connection weights after one complete cycle of pattern representations is proportional to this derivative and hence, that the backpropagation algorithm implements a steepest descent in E. Of course, it is strictly true only if the values of the connection weights are not changed during this cycle.

2.2.3 Algorithm: Backpropagation Method

The backpropagation algorithm can be summarized as in the following steps:

Step 1. Set the initial values of w_{kj}, w_{ji}, θ_j, θ_i, η, and α.

Step 2. Apply the input to the neural network, specify the corresponding desired output τ_k, and calculate O_j,O_k, and δ_k by the formula

$$\delta_k = O_k(1 - O_k)(\tau_k - O_k). \tag{2.2.33}$$

Step 3. Change the connection weights by

$$\Delta w_{kj}(t+1) = \eta \delta_k O_j + \alpha \Delta w_{kj}(t). \tag{2.2.34}$$

Step 4. Calculate δ_j by

$$\delta_j = O_j(1 - O_j) \sum_k \delta_k w_{kj}. \tag{2.2.35}$$

Step 5. Change the connection weights by

$$\Delta w_{ji}(t+1) = \eta \delta_j O_i + \alpha \Delta w_{ji}(t). \tag{2.2.36}$$

Step 6. $t \rightarrow t+1$ and go to Step 2.

2.2.4. Some Discussions on the Backpropagation Algorithm

The backpropagation is based on the steepest descent (gradient) algorithm to find the minimum of the error which means to find the local minimum or plateau as shown in Fig. 2.2.3. Ideally, we seek a global minimum which is the lowest bottom of the valley in the entire terrain. Since a local minimum is surrounded by higher ground, and the network usually does not leave a local minimum by the standard backpropagation algorithm stated above if we encounter the local minimum.

The appearance of a local minimum is not always a significant problem. We can expect that the backpropagation networks converge to a good mean squared value when the training examples are clearly distinguishable. In order to avoid entering the local minimum, we may change the learning parameters, the number of hidden nodes, or initial values of connecting weights. Basically, these techniques are trying to change the scenario involved with moving about on the mountain terrain such that the neural network can avoid a local minimum. Recently, the genetic algorithm has been used to find a better set of these parameters.

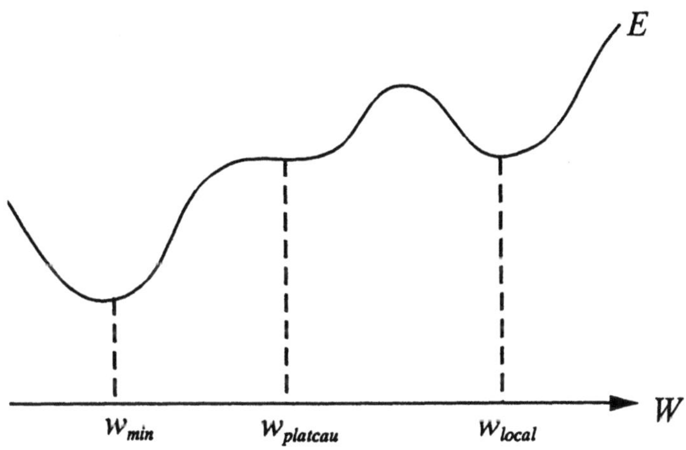

Fig. 2.2.3. Local minimum in the backpropagation algorithm.

We have selected a sigmoid function of (2.2.5) to get the back-propagation algorithm given in Section 2.2.2. It is not necessary to limit the activation function to the type. If we can select another type like

$$f(x) = \frac{1}{1+e^{-\frac{x}{T}}}, \tag{2.2.37}$$

then we get

$$f'(x) = \frac{1}{T} f(x)(1 - f(x)). \tag{2.2.38}$$

Thus, we have only to replace δ_k and δ_j of (2.2.33) and (2.2.35) to $\frac{\delta_k}{T}$ and $\frac{\delta_j}{T}$, respectively. If we choose small T, then δ_k and δ_j become large and hence, changing rate of the connection weights becomes large. But if we choose large T, then δ_k and δ_j become small and the changing rate of the connection weights becomes small. Therefore, if we change T during the learning process, we can control the convergence speed of the back-propagation algorithm.

It should be noted that in a practical learning situation in which the desired outputs are binary {0, 1}, the system can never actually achieve these values except when x/T goes to infinity. Thus, if we select smaller value of T, the output of the neural network tends to converge faster. In the case where T=1, the derivatives O_k $(1-O_k)$ and O_j $(1- O_j)$ reach their maximum values when $O_k = O_j = 0.5$. Since $0 \le O_k, O_j \le 1$, their minima have been attained when they approach zero or one. Since the amount of change in the given connection weights is proportional to the derivatives, the connection weights will be changed most for those units that are near their midrange.

Another feature of the activation function is that the neural network cannot reach its extreme value of 1 or 0 without infinitely large connection weights. Thus, in a practical learning situation when the desired outputs are binary {0, 1}, the neural network can never achieve these values. Therefore, we may regard the values of 0.1 and 0.9 as the desired values instead of 0 and 1, respectively even though we are seeking the desired values of {0, 1}.

The learning rate constant η is proportional constant to the change of connection weight to the gradient of the error E with respect to the weight. The larger this constant, the larger the changes in the connection weights. Usually, a learning rate η is selected as large as possible without leading to oscillation. Since we have included the momentum term to prevent from leading to oscillation, we must choose a suitable value of α. The momentum term filters out high-frequency variations of the error-surface in the connection weight space. This is useful in the connection weight spaces containing long ravines with sharp curvature across the ravine. Since the sharp curvature tends to cause divergent oscillations across the ravine, it is necessary to take small step size by selecting small learning rate. Using the momentum term, we can select the effective learning rate. Rumelhart *et al.* [11] have recommended the value of α is about 0.9.

The other problem to be addressed is to determine the size of the hidden nodes or hidden layers. From regression analysis, it is well-known that if we choose many kind of explanation variables, then the prediction error becomes small. For example, assume that p kinds of explanation variables $x_1, x_2 ..., x_p$ and one objective variable y are chosen to estimate the following linear regression

$$y_k = a_0 + a_1 x_{1k} + \ldots + a_p x_{pk} + e_k \tag{2.2.39}$$

where k = 1, 2, ..., N.

If we select N = p, then the prediction errors e_k (k = 1, 2, ..., p) can be equal to zero. It seems that perfect prediction can be achieved by the regression equation. For practical purposes, it cannot be used since N should be larger than p. Therefore, we must be careful to note that minimizing the squared error by increasing the number of neurons does not mean building a good neural network. Therefore, if we take a large number of neurons, the neural networks may learn the noisy data perfectly. In this case, we select a neural network of small size even though the error is not so small as shown in Fig. 2.2.4. In this figure, $\hat{E}$ denotes the error curve obtained by perfect learning the observation data denoted by and E^o denotes the other error curve which produces smaller prediction errors for unseen data denoted by ◘. This phenomenon is called the overfitting problem. To overcome this problem, there have been proposed some techniques to reduce the neural network size.

2.3 Conclusions

In this chapter, simplified biological neural network structures and functions have been briefly described. It can be understood that much of the artificial neural network terminology, model, and functions have been borrowed from biological neural networks. However, learning in artificial neural networks does not resemble much to learning in biological neural networks. Artificial neural networks are trained through simplified learning algorithms that adapt the synaptic weights which can be classified as either supervised or unsupervised. In this chapter, we have described and derived the backpropagation algorithm which is one of the most widely used algorithms for mulilayered neural networks. This algorithm forms the basis of the neuro-control schemes discussed in this book.

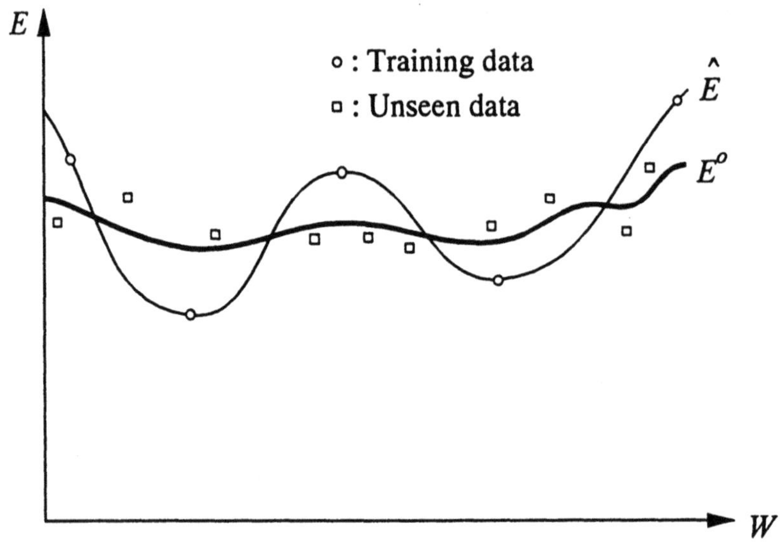

Fig. 2.2.4. Overfitting phenomenon.

2.4 References

[1] Cajal, S.R., "Histologie du systeme nerveux de l-homme et des vertebres", Maloine, Paris, 1909.

[2] Dayhoff, J.E., *"Neural Network Architectures: An Introduction"*, Van Nostrand Reinhold, New York, 1990.

[3] McCulloch, W.S. and W. Pitts, "A logical calculus of the ideas immanent in nervous activity", *Bulletin of Mathematical Biophysics*, Vol. 9, pp.127-147, 1943.

[4] Hebb, D.O., *"The Organization of Behavior"*, Wiley, New York, 1949.

[5] Bernstein, N., "Profiles:AI, Marvin Minsky", The New Yorker, pp.50-126, 1981.

[6] Rosenblatt, F., *"Principles of Neurodynamics: Perceptrons and the Theory of Brain Mechanisms"*, Spartan Book, Washinghton D.C., 1961.

[7] Minsky, M.L. and S.A. Papert, *"Perceptrons"*, MIT Press, Cambridge, MA, 1969.

[8] Hopfield, J.J., "Neural networks and physical systems with emergent computational abilities", *Proc. of the National Academy of Sciences*", Vol. 79, pp. 2554-2558, 1982.

[9] Hopfield, J.J. and D.W. Tank, "Neural computation of decisions in optimization problems", *Biological Cybernetics*, Vol. 52, pp. 141-152, 1985.

[10] Werbos, P.J., "Beyond regression: New tools for prediction and analysis in the behavioral sciences", Ph.D. Thesis, Applied Mathematics, Harvard University, Nov., 1974.

[11] Rumelhart, D.E. *et al.*, *"Parallel Distributed Processing: Explorations in the Micro Sructure of Cognition Vol.I: Foundations"*, MIT Press, Cambridge, MA, 1986.

[12] Rumelhart, D.E. and D. Zipser, "Feature discovery and competitive learning", Cognitive Science, Vol. 9, pp. 75-112, 1985.

[13] Grossberg, S., "Adaptive pattern classification and universal recoding: I. Parallel development and coding of neural feature detectors", *Biological Cybernetics*, Vol. 23, pp. 121-134, 1976.

[14] Kohonen, T., "Self-organized formation of topologically correct feature maps", *Biological Cybernetics*, Vol. 43, pp. 59-69, 1982.

[15] Widrow, B., "ADALINE and MADALINE - 1963", Plenary Speech, Vol. I, *Proc. of IEEE First Int. Conf. on Neural Networks*, San Diego, CA, pp. 143-158, 1987.

CHAPTER 3
TRADITIONAL CONTROL SCHEMES

3.1 Introduction

In this chapter we are going to discuss several other methods of control. These methods include PID (proportional-plus-integral-plus derivative) control, self-tuning control, self-tuning PID control, generalized predictive control, and also fuzzy logic control. One of the earliest controllers that were used for control were the PI and PID controllers. PI and PID controllers have been proven to be remarkably effective in regulating a wide range of processes. The use of PI and PID controllers does not require an exact process model and hence, they are effective on industrial processes whose models are considerably difficult to derive. The PI and PID controllers are based on classical control theory and much easier to understand. Field engineers and process operators are able to relate the parameter settings and control system actions.

Moreover, several methods for tuning the PI and PID controllers have been introduced over the last few decades. Ziegler and Nichols [1] proposed a tuning method based on step response data which has been widely used over the years and has been proven to be successful to some extent. Takahashi *et al.* [2] introduced a tuning method for discrete time PI and PID controllers based on the Ziegler-Nichols design. A more recent method for auto-tuning PI and PID controllers was proposed by Nishikawa *et al.* [3]. This method requires an input test signal to be applied to estimate the process parameters. The optimal settings of the PI and PID parameters are obtained from the minimization of the weighted Integral of Squared Error (ISE). Another method of auto-tuning the PID controller was suggested by Hagglund and Åström [4]. This method is essentially based on the Ziegler-Nichols closed loop formula suggested in the seminal paper [1].

However, in spite of the advantages of PI and PID controllers outlined above, there remain several drawbacks. For example, if the operating point of the process is changed due to disturbances, the controller parameters need to be adjusted manually to obtain another optimal settings. This procedure known as tuning needs to be performed by a skilled operator. For systems with interacting loops, this process could be difficult and time consuming. Furthermore, for processes with variable time-delays, varying plant parameters, large nonlinearities and innegligible process noise, the use of PI and PID controllers may not result in optimum performance. In addition to these, the tuning methods for the PI and PID controllers have some drawbacks. For example, the Ziegler-Nichols [1] method is sensitive to disturbances due to its reliance on open loop experiments. And the tuning method proposed by Nishikawa *et al.* [3] requires man-machine interaction in which the operator needs to generate input signals every time the parameters have to be modified in order to adapt to changes in the process dynamics.

Due to these reasons, the idea of self-tuning control (STC), in which the parameters of the controller can be adjusted on-line, was introduced by Kalman [5]. He postulated building a machine which adjusts itself automatically to control an arbitrary dynamic process. The idea was important in the evolution of the self-tuning regulator. However, the technique was not actively pursued due to the inadequacy of the theory and technology developed. The revival of the theory which initiated a great interest among control engineers and researchers alike, was through a paper by Åström and Wittenmark [6]. They have uncovered a striking fact that in some cases, a combination of a simple least squares estimation and minimum variance control yields a self-tuning property. The paper also led to a series of practical applications, which themselves engendered new theoretical development as the problems involved in implementimg self-tuning control needed to be clarified. In the seventies and eighties, the growth in the digital computers technology has been tremendously rapid and this played an important role in increasing the interest in the field of self-tuning control both in theory and practice.

Research and development of self-tuning algorithms have been carried out both in terms of expanding existing schemes into new applications areas and also by the introduction of new approaches to the regulatory and servo problems dealt with by available controllers. Another development of research in this area is on self-tuning PI (STPI) or PID (STPID) controllers. The main idea of these types of controllers is to combine the ability of self-tuning controller which can accommodate for changes in the plant parameters on-line and the simplicity of the PID controllers structures.

There are several schemes of STPID controllers proposed in the literature. For example, Wittenmark [7] proposed STPID controller based on pole placement design. In this design, a PID algorithm is fitted into the control structure which is calculated via pole placement design. The method has some restrictions in that the order of the process which can be controlled is limited. A modification of the method which allows a more general form of processes to be controlled was proposed by Savelli *et al.* [8]. Later, Gawthrop [9] proposed a STPI or STPID controller for continuous time systems. The PI or PID algorithm is automatically derived from reasonable assumptions about the dynamics of the controlled process and also suitable modeling of the non-zero mean disturbances. Other forms of STPID can be found in Banyasz and Kevicsky [10], Ortega and Kelly [11], Radke and Isermann [12], Jones and Porter [13], Proudfoot *et al.* [14], Yamamoto *et al.* [15], Rad and Gawthrop [16], etc.

An alternative STPI or STPID controller which is of interest in this book was introduced by Cameron and Seborg [17]. The algorithm is based on the generalized minimum variance control (Clarke and Gawthrop [18], [19]). The STC law is orientated to have a PI or PID structure. This is done by making reasonable assumptions regarding the user defined polynomial functions of the STC. An integral action which is required to eliminate disturbance of step-wise nature and steady-state error is introduced in the algorithm in an alternative manner. The controller parameters are obtained using a parameter estimation scheme. For a PI structure, the number of controller parameters needed is two, whilst for a PID structure, the required number of controller parameters is three. This requirement places some limitation in the order of the poles of the open loop systems to be controlled but no limitation is placed on the order of the zeros. However, it can be argued that, for a higher order system, model order reduction technique can be applied.

A multivariable approach of the STPID controller will also be discussed in Section 3.3. The formulation of multivariable STPID controllers takes into consideration the same control problems as their SISO counterparts. These problems are usually nonlinearities, varying plant dynamics, deterministic and stochastic disturbances, and time-delays. In multivariable case, there are also some technical difficulties which do not appear in SISO controllers such as the problems arising from the well known topics of multivariable control theory. For example, in multivariable case, the input-output relationship has to be represented in the form of left or right matrix fraction description. Also, the polynomial matrices do not commute and to solve this problem some additional matrices have to be introduced. This results in an increase in the formulations and calculations which are algebraic in nature (Johansson [20]). Certain issues concerning parameterization, unequal time-delays, and stability also posed some problems in the field of multivariable STPI or STPID controller.

Another recent development in the field of adaptive control is the long range prediction control method. These methods are based on the principle shown in Fig. 3.1.1. The strategy of the method is that at each present time t, a forecast of the process output over a long-range time horizon is made. This forecast is made implicitly or explicitly in the control algorithm and is based on the mathematical model of the process dynamics.

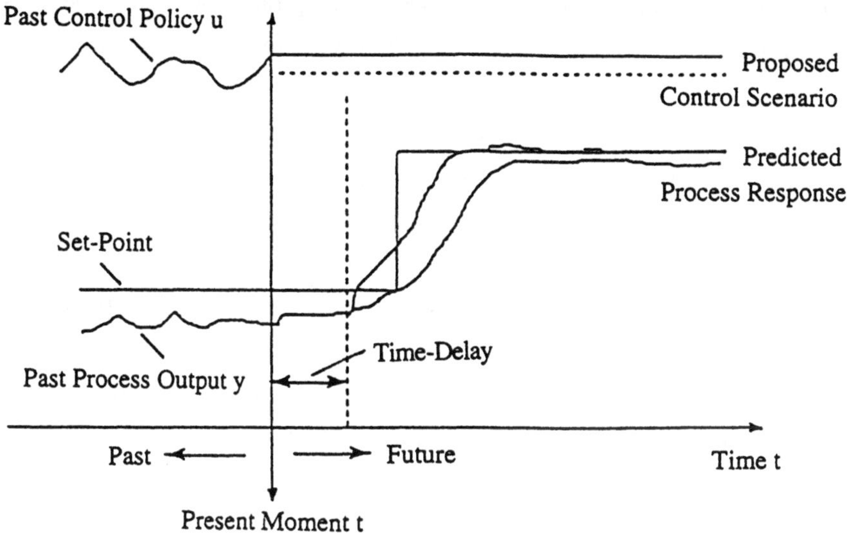

Fig. 3.1.1. Long range prediction strategy.

Moreover, it is a function of the future control scenario which is being presented as being constant over the prediction horizon. From the several control scenarios, the strategy which brings the predicted process output back to the set-point in the best way according to the specific control objective is selected. The resulting best candidate is then applied as a control action to the real process input but only at the present time. At the next sampling instant, the whole procedure is repeated, leading to an updated control action with corrections based on the latest measurements. This strategy is also known as receding horizon strategy.

Several of the control methods based on long range predictive control have been reported, some of which are MAC (Model Algorithmic Control) reported by Richalet *et al.* [21], DMC (Dynamic Matrix Control) by Cutler and Ramaker [22], EPSAC (Extended Predictive Self Adaptive Control) by

De Keyser and Van Cauwenberghe [23], EHAC (Extended Horizon Adaptive Control) by Ydstie [24], Predictor based self-tuner by Peterka [25], and GPC (Generalized Predictive Control) by Clarke *et al.* [26], [27], [28]. A rather comprehensive comparative study of various long range predictive methods describing the relative merits of the method was presented by De Keyser *et al.* [29].

Earlier long range predictive methods such as MAC and DMC use deterministic impulse or step response model. These models require graphical or reaction curve experiments which are rather laborious procedures. GPC algorithm uses a CARIMA (Controlled Auto-Regressive and Integrated Moving Average) model whose parameters can be obtained via parameter estimation scheme such as recursive least squares estimator. The range of the tuning parameters of the GPC is also more general and hence, allows more flexibility in the design of the closed loop system as compared to the other methods such as EPSAC or EHAC which can be considered as subsets of GPC.

One of the most recent successful control applications is fuzzy logic control. Fuzzy logic theory was first developed by Zadeh in the mid-sixties [30] and currently it has been successfully applied to many industrial systems and domestic products. In the final section we will discuss fuzzy set theory and fuzzy logic control.

3.2 Discrete-Time PI and PID Controllers

The discrete PI or PID controller form which has been adopted for this algorithm is the velocity form. The reason for choosing this type of PI or PID controllers is due to the many advantages of these controllers over the other forms of PI or PID controllers. The advantages of the velocity form of PI or PID controllers include the ability to cope with sudden large changes in the set-point and when the operation is switched from manual to automatic, initialization is not needed. Therefore, if there are sudden large changes in the set-point or at the start of process operation, it exhibits a better response than other types of PI or PID controller. It is also useful in preventing integrator saturation which usually occurs if the output saturates and the controller continues integrating the error. A velocity form of a discrete PID control scheme can be written as follows:

$$\Delta u(t) = k_c((e'(t) - e'(t-1)) + \frac{T_s}{2T_i}(e'(t) + e'(t-1)) + \frac{T_d}{T_s}(e'(t) - 2e'(t-1) + e'(t-2))). \qquad (3.2.1)$$

Equation (3.1) can be rearranged to be in the following form:

$$\Delta u(t) = K_P(e'(t) - e'(t-1)) + K_I e'(t) + K_D(e'(t) - 2e'(t-1) + e'(t-2)) \qquad (3.2.2)$$

where $K_P = k_c - \frac{1}{2}K_I$, $K_I = \frac{k_c T_s}{T_i}$, $K_D = \frac{k_c T_d}{T_s}$, and $\Delta = 1 - z^{-1}$.

Now, k_c, T_i , T_d and T_s are the gain, reset time, derivative time, and sampling time, respectively. K_P, K_I, and K_D are the proportional gain, integral gain, and derivative gain, respectively. The error signal e'(t) can be written as

$$e'(t) = r(t) - y(t) \qquad (3.2.3)$$

where r(t) is the desired set-point. We now have

$$\Delta u(t) = K_P(r(t) - r(t-1) - y(t) + y(t-1)) + K_I(r(t) - y(t)) + K_D(r(t) - 2r(t-1) + r(t-2) - y(t) + 2y(t-1) - y(t-2)). \qquad (3.2.4)$$

A sudden large change in the set-point would result in the proportional and derivative control action producing a large change in the controller output. To suppress this phenomenon, the set-point r(t) is assumed to be constant for a while until the next step change takes place. Now (3.2.4) becomes

$$\Delta u(t) = K_I r(t) - [K_P + K_I + K_D] y(t) + [K_P + 2K_D] y(t-1) - K_D y(t-2). \quad (3.2.5)$$

3.3 Self-Tuning Control

Self-tuning control can be thought of as performing two basic tasks in a feedback loop. A general configuration of the self-tuning control method is depicted in Fig 3.3.1. The configuration represents two important tasks of a self-tuner. The first task is information gathering of the present process behaviour. This involves continuous determination of the actual condition of the process to be controlled based on the measurable process input, process output, and state signals.

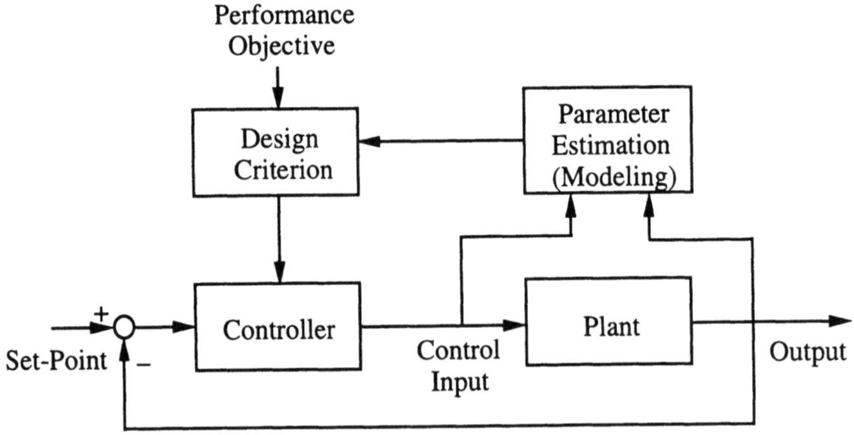

Fig. 3.3.1. A general configuration of self-tuning controller.

The information is used for system identification which involves the determination of model structure, parameter estimation, and also signal estimation of non-measurable signals such as noise signals in stochastic systems. The determination of the model structure requires a construction of a mathematical representation of the system relevant to the problem at hand.

Parameter estimation is a key element in a self-tuner and is performed on-line. A number of recursive parameter estimation schemes are employed for self-tuning control. The most popular scheme is the recursive least squares estimatation method [31], [32] or the more robust extension of this method, the UD factorization method [33].

The second task of a self-tuner is the design task which is usually based on control performance criterion optimization. A performance objective is defined for a particular system and some decisions have to be made as to how the controller will be adjusted or adapted. Based on these decisions, a new set of controller parameters are calculated replacing the old parameters in the control loop to meet the performance objective. This part of the process is known as the validation phase and is one of the most favourable properties of the self-tuner since it is performed on-line and in real-time. In conventional method, the validation process is usually performed off-line and often proves unsatisfactory and sometimes results in repetition of the whole modelling and design process. The calculation of the control law is based on a procedure called certainty equivalent in which the current parameter estimates are accepted while ignoring their uncertainties.

The approach of estimating the parameters of the transfer function of the process and the disturbance is termed as indirect self-tuning or explicit method. An example of this type of method is the self-tuning regulator of Åström and Wittenmark [6] also known as minimum variance control. The criterion of the regulator is to minimize the variance of the output y(t) at each sampling instant t. This can be achieved by first predicting the future output d step ahead, y(t+d) and then setting the current control u(t) to make this output zero. However, this method makes no attempt to ensure that the set-points are followed optimally and may produce an excessive control inputs. Another disadvantage is that the closed loop is unstable due to the cancellation of the process zeros if the process is non-minimum phase. To overcome these problems, Clarke and Gawthrop [18], [19] modified the cost criterion into minimizing not only the output variance, but also the function which includes the control input, output, and also set-point variations to allow a wider variety of performance objectives. The criterion penalizes the control signal to reduce the excessive peaking of the control signals and provides the means for optimal set-point following. In addition, the modified criterion enables stable control of non-minimum phase processes. This type of self-tuning controller is also known as GMV (generalized minimum variance control). Various forms of extensions and refinements of this type of self-tuning controller have appeared in the literature in recent years, see for example, Gawthrop [34] and Clarke [35]. Clarke *et al.* [36], Tuffs and Clarke [37], Clarke [38], and Clarke *et al.* [39].

For derivation of this algorithm, a system in the form of a CARMA (controlled auto-regressive moving average) model is considered. The system can be modeled in mathematical form as follows:

$$A(z^{-1})y(t) = B(z^{-1})u(t-d) + C(z^{-1})\xi(t) \qquad (3.3.1)$$

where y(t) is the measured output, u(t) is the control input, $\xi(t)$ is an uncorrelated sequence of random variables with zero mean and covariance σ, d is the time-delay, and t is the time in sample instant (integer). $A(z^{-1})$, $B(z^{-1})$, and $C(z^{-1})$ are expressed in terms of the backward shift operator z^{-1}

$$A(z^{-1}) = 1 + a_1 z^{-1} + a_2 z^{-2} + \ldots\ldots. + a_{n_a} z^{-n_a} \qquad (3.3.2)$$

$$B(z^{-1}) = b_0 + b_1 z^{-1} + b_2 z^{-2} + \ldots\ldots. + b_{n_b} z^{-n_b} \qquad (3.3.3)$$

$$C(z^{-1}) = 1 + c_1 z^{-1} + c_2 z^{-2} + \ldots\ldots. + c_{n_c} z^{-n_c}. \qquad (3.3.4)$$

It is assumed that the roots of $C(z^{-1})$ lie within the unit disc. No assumption is made concerning the roots of $A(z^{-1})$ and $B(z^{-1})$ polynomials, i.e., the roots of $A(z^{-1})$ may be outside the unit disc in which the plant is open loop unstable, or the roots of $B(z^{-1})$ may be outside the unit circle which means that the plant is non-minimum phase.

The cost criterion adopted in this algorithm is of the following form

$$J = E\left[(P(z^{-1})y(t+d) - R(z^{-1})r(t))^2 + (Q(z^{-1})u(t))^2\right] \tag{3.3.5}$$

where E[.] denotes the expectation value, $P(z^{-1})$, $R(z^{-1})$, and $Q(z^{-1})$ are user defined polynomials and r(t) is the set-point. $P(z^{-1})$ is a rational transfer function of the form

$$P(z^{-1}) = \frac{P_n(z^{-1})}{P_d(z^{-1})}. \tag{3.3.6}$$

Here, $P_n(z^{-1})$ and $P_d(z^{-1})$ are polynomials with degree np_n and np_d, respectively. The cost criterion differs in some respects to the cost criterion of minimum variance control proposed by Åström. The minimum variance control only concentrates on the regulatory control of the process and that is regulating the system output y(t) about zero against random disturbances. However, in many industrial situations, it is required to regulate the system output against random disturbance at a certain constant level. The GMV controller takes into consideration both the regulatory control and also the servo control. In addition to that, the GMV controller has a costing on the control input which prevents excessive control input being produced in the occurrence of some cancellation of zeros which are outside the stability region. The approach introduces the pseudo-output $\phi(t)$ defined by

$$\phi(t) = P(z^{-1})y(t)+Q(z^{-1})u(t-d)-R(z^{-1})r(t-d). \tag{3.3.7}$$

The system can now be considered as a generalized system output with a feed-forward term, a filtering action on the output, and the set-point. Many self-tuning strategies are based on predictive control design and the prediction horizon is the time-delay d. An optimal prediction of the output at time t+d can be obtained at time t if the input is chosen such that the disturbances are neutralized. To obtain an optimal prediction of the pseudo-output, we need to consider first the identity equation of the form:

$$C(z^{-1})P_n(z^{-1})=E(z^{-1})A(z^{-1})P_d(z^{-1})+z^{-d}F(z^{-1}) \qquad (3.3.8)$$

where

$$E(z^{-1}) = 1+e_1z^{-1}+\dots\dots\dots + e_{d-1}z^{-(d-1)} \qquad (3.3.9)$$

$$F(z^{-1}) = f_0 + f_1z^{-1} + \dots\dots f_{n_f}z^{-(n_f)} \qquad (3.3.10)$$

and

$$n_f = \max(n_a + n_{Pd}-1,\ n_c+n_{Pn}-d). \qquad (3.3.11)$$

Multiplying the system equation (3.3.1) by $E(z^{-1})$, we get

$$E(z^{-1})A(z^{-1})y(t+d) = E(z^{-1})B(z^{-1})u(t)+E(z^{-1})C(z^{-1})\xi(t+d). \qquad (3.3.12)$$

Substituting for $E(z^{-1})A(z^{-1})$ from (3.3.8) in (3.3.1) results in

$$P(z^{-1})y(t+d) = F(z^{-1})y_f(t) + E(z^{-1})B(z^{-1})u(t) + E(z^{-1})C(z^{-1})\xi(t+d). \qquad (3.3.13)$$

Here, $y_f(t) = P_d(z^{-1})^{-1}y(t)$. Adding $Q(z^{-1})C(z^{-1})u(t) - C(z^{-1})R(z^{-1})r(t)$ to both sides of (3.3.13) gives

$$\begin{aligned} C(z^{-1})(P(z^{-1})y(t) + Q(z^{-1})u(t-d) - R(z^{-1})r(t-d)) \\ = F(z^{-1})y_f(t) + (C(z^{-1})Q(z^{-1}) + E(z^{-1})B(z^{-1}))u(t) \\ + E(z^{-1})C(z^{-1})\xi(t+d) - C(z^{-1})R(z^{-1})r(t) \end{aligned} \qquad (3.3.14)$$

which is equivalent to

$$C(z^{-1})\phi(t+d) = F(z^{-1})y_f(t) + (C(z^{-1})Q(z^{-1}) + E(z^{-1})B(z^{-1}))u(t) + E(z^{-1})C(z^{-1})\xi(t+d) - C(z^{-1})R(z^{-1})r(t). \quad (3.3.15)$$

If we define $\phi^*(t+d/t)$ as the optimum prediction of $\phi(t+d)$ based on the measurement up to time t, $\tilde{\phi}(t+d)$ as the prediction error, and assuming $C(z^{-1})=1$, we have

$$\phi^*(t+d/t) = F(z^{-1})y_f(t)+(G(z^{-1})+Q(z^{-1}))u(t)-R(z^{-1})r(t) \quad (3.3.16)$$

and

$$\tilde{\phi}(t+d) = \phi(t+d) - \phi^*(t+d/t) = E(z^{-1})\xi(t+d) \quad (3.3.17)$$

Here, $G(z^{-1})=E(z^{-1})B(z^{-1})$. Noting that the future errors in $\xi(t+1)$, ... and $\xi(t+d)$ are uncorrelated with the present and past input and output measurements, the minimum variance control of $\phi(t+d)$ is obtained by choosing u(t) such that $\phi^*(t+d/t)$ is set to zero. We now have

$$u(t) = \frac{R(z^{-1})r(t) - F(z^{-1})y_f(t)}{G(z^{-1}) + Q(z^{-1})}. \quad (3.3.18)$$

Since at time t, $Q(z^{-1})u(t)$ and $R(z^{-1})r(t)$ are known, the prediction of $\phi(t+d)$ can be reduced to

$$\phi_y(t) = P(z^{-1})y(t). \quad (3.3.19)$$

In the sequel, the optimal d step ahead prediction of $\phi_y(t)$ is

$$\phi_y^*(t+d) = F(z^{-1})y_f(t) + G(z^{-1})u(t). \quad (3.3.20)$$

The controller parameters $F(z^{-1})$ and $G(z^{-1})$ can be easily obtained from the identity (3.13) if $A(z^{-1})$ and $B(z^{-1})$ are known. However, if $A(z^{-1})$ and $B(z^{-1})$ are unknown, then recursive least squares method discussed in the following section can be used to estimate the elements of $F(z^{-1})$ and $G(z^{-1})$. The vectors for the parameters and data regression are as follows:

$$\hat{\theta}(t) = [\hat{f}_0(t), \hat{f}_1(t), \ldots, \hat{f}_{nf}(t), \hat{g}_0(t), \hat{g}_1(t), \ldots, \hat{g}_{ng}] \tag{3.3.21a}$$

$$x^T(t) = [y_f(t-1), y_f(t-2), \ldots, u(t-1), u(t-2), \ldots] \tag{3.3.21b}$$

where $\hat{f}_0, \hat{f}_1, \ldots$ and $\hat{g}_0, \hat{g}_1, \ldots$ denote the estimate of the elements of $F(z^{-1})$ and $G(z^{-1})$, respectively.

For self-tuning controller, the parameter estimation is performed recursively and the observations are obtained sequentially. A recursive parameter estimation can be visualized according to Fig. 3.3.2. The estimate of the current output $\hat{y}(t)$ is obtained based on the past information of the model provided by $\hat{\theta}(t-1)$. The modeling error $\varepsilon(t)$ is generated by comparing the estimated output with the observed output. The error is then minimized in the least squared sense and then $\hat{\theta}(t-1)$ is updated to a new value $\hat{\theta}(t)$. The estimates of $\hat{\theta}(t)$ can be obtained by means of three important recursive equations as follows:

$$\hat{\theta}(t) = \hat{\theta}(t-1) + K(t)[y(t) - x^T(t)\hat{\theta}(t-1)] \tag{3.3.22}$$

$$K(t) = P(t-1)x(t)[1 + x^T(t)P(t-1)x(t)]^{-1} \tag{3.3.23}$$

$$P(t) = (I - K(t)x^T(t))P(t-1) \tag{3.3.24}$$

where P(t) known as the covariance matrix is defined as

$$P(t) = [x^T(t)x(t)]^{-1}. \tag{3.3.25}$$

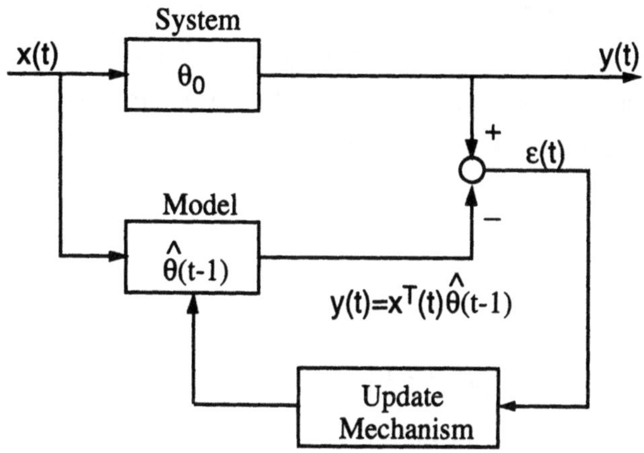

Fig. 3.3.2. Block diagram of a recursive parameter estimation scheme.

3.4 Self-Tuning PI and PID Controllers

The dominance of PI and PID controllers in the process control industries is the main driving force towards the development in the theories and practice of STPI and STPID controllers. Although self-tuning controller is more flexible and provides a more systematic way of dealing with uncertainties, nonlinearities, and time varying plant parameters, its applications in the process control industries are still not very encouraging.

Self-tuning regulators can be viewed as performance oriented in that, the closed loop performance is specified by the user and the algorithm sets out to attain this performance even though the plant parameters or the drifts are unknown. Indeed, this implies that the desired performance of the plant can only be achieved given the saturation characteristics of the control actuator and the skill of the plant engineers is crucially important. A progressive step is to provide the plant engineers or users a greater intuitive understanding of the ultimate closed loop performance under self-tuning control. In this light, the theory of self-tuning controller is moving rapidly towards broadening the range of possible control objectives and interpreting them in classical control engineering terms which are easier to understand.

The derivation of the STPI or STPID controller of Cameron and Seborg [17] can be obtained from the combination of (3.2.5) and (3.3.19). In order for the control law of (3.3.19) to have a PID structure of (3.2.5), $F(z^{-1})$ must be specified to be a second order polynomial. Now the order of $F(z^{-1})$ is given by

$$n_f = n_a + n_{pd} - 1 \tag{3.4.1}$$

assuming that $n_a + n_{pd} - 1 > n_{pn} + n_c$. For a first order system, a PID structure is obtained if we choose a second order prefilter polynomial $P_d(z^{-1})$. Now since $Q(z^{-1})$ can be chosen by the user, integral action can be introduced by letting

$$\frac{\Delta}{v} = G(z^{-1}) + Q(z^{-1}) \tag{3.4.2}$$

where v is a constant.

The steady-state error can be eliminated by letting

$$R(z^{-1}) = H_0 = \sum_{i=0}^{n_f} \frac{F_i(z^{-1})}{P_d(z^{-1})} \Big| z^{-1} = 1. \tag{3.4.3}$$

Using (3.4.2) and (3.4.3), (3.3.18) can be written as

$$\Delta u(t) = v(H_0 r(t) - (\hat{f}_0 + \hat{f}_1 z^{-1} + \hat{f}_2 z^{-2}) y_f(t)) \tag{3.4.4}$$

which is the control law for a self-tuning controller with a PID structure. Expressions for the corresponding PID controller parameters are as follows:

$$K_I = \frac{-v(\hat{f}_0+\hat{f}_1+\hat{f}_2)}{P_d(1)}, \tag{3.4.5}$$

$$K_P = \frac{-v(\hat{f}_1+2\hat{f}_2)}{P_d(1)}, \tag{3.4.6}$$

and

$$K_D = \frac{-v\hat{f}_2}{P_d(1)}. \tag{3.4.7}$$

It is interesting to note that from the corresponding PID controller expressions, v is expected to give the same effect to the STPID controller as Kc does to the PID controller. A small value of v would result in a more oscillatory control. A reasonable choice for the filter $P(z^{-1})$ is a lead network. This implies that the closed loop is a low pass filter. The choice of v, $P_n(z^{-1})$, and $P_d(z^{-1})$ is more on a trial and error basis, but a reasonable approach in selecting these parameters is to select a suitable value of v so that the closed loop response is stable and not too oscillatory and then to vary the values of $P_n(z^{-1})$ and $P_d(z^{-1})$ to obtain a reasonable performance for the application.

The software implementation of the STPID controller at each sampling instant can be summarized as follows:

<u>Computation of STPI controller</u>

Step 1. Set the reference input.

Step 2. Read the input and output data to the microcomputer.

Step 3. The prefilters coefficient $P_n(z^{-1})$ and $P_d(z^{-1})$ are pre-determined.

Step 4. Recursive least squares algorithm is performed using (3.3.22), (3.3.23), and (3.3.24)

Step 5. The control input can be calculated using (3.4.4).

Step 6. The corresponding PID parameters can be calculated from (3.4.5) to (3.4.7).

3.4.1 Closed Loop System

The closed loop expression for the STPI or STPID controller is as follows:

$$(\Delta A(z^{-1}) - z^{-d}B(z^{-1})\frac{F(z^{-1})}{P_d(z^{-1})})y(t) = z^{-d}vB(z^{-1})H_0 r(t) + \Delta\xi(t) \tag{3.4.8}$$

where $\xi(t)$ is the disturbance acting on the process. It is obvious that the choice of the prefilter polynomials $P_n(z^{-1})$ and $P_d(z^{-1})$ will have some effect on the closed loop system. It can also be seen that perfect asymptotic tracking is achieved for constant reference signal if

$$H_0 = \frac{F(1)}{P_d(1)} \tag{3.4.9}$$

which conforms with (3.4.3). Also, note that at steady-state, the disturbance term equals zero, implying that the closed loop system is able to regulate the effect of load disturbance to zero.

3.4.2 Some Interpretations Based on a Simulation Example

In most control system designs, some tuning parameters have to be preselected. In the case of STPI or STPID controller, the tuning parameters are v, $P_n(z^{-1})$, and $P_d(z^{-1})$. The tuning parameters of the STPI or STPID controller play an important role in shaping up the closed loop response of the system under control as it is related to the controller parameters. For PI or PID controllers, the parameters K_P, K_I, and K_D have to be selected, although they remain unchanged unless some changes occur in the plant in which retuning of the parameters is required.

In order to assist in the selection of the tuning parameters for STPI or STPID such as v, $P_n(z^{-1})$, and $P_d(z^{-1})$, some simulation results are shown in this section. Some study of the properties and the role of the tuning parameters of STPI or STPID is made through the simulation examples. We also perform some simulation examples using PI or PID controller whose parameters are fixed to make some comparisons with the STPI or STPID controller. The simulation examples are carried out for a total of 300 sampling instants. The least squares estimate is used to estimate the parameters and the covariance matrix is initially chosen to be ten multiplied by the unit matrix. The forgetting factor λ is chosen to be one.

A second order continuous time system with time-delay is considered in this example. The system has a transfer function given as follows:

$$Y(s) = \frac{e^{-s}}{(s+0.1)(s+1.0)}U(s).$$

A discrete time form of the transfer function can be obtained by means of Z-transform and zero order hold with a sampling time of 0.5 sec. The discrete time transfer function is given by

$$GH(z) = \frac{z^{-3}(0.105+0.087z^{-1})}{1-1.557z^{-1}+0.576z^{-2}}.$$

In time domain, the discrete time transfer function can be written as follows:

$$y(t)=1.557y(t-1)-0.576y(t-2)+0.105u(t-3)+0.087u(t-4).$$

For this system, a STPID controller is realizable only if $P_d(z^{-1})$ is chosen as a first order polynomial. No restriction on the order of $P_n(z^{-1})$ is required and we choose $P_n(z^{-1})$ to be a second order polynomial. Therefore,

$$P_d(z^{-1}) = 1+p_{d1}z^{-1},$$

and

$$P_n(z^{-1}) = 1+p_{n1}z^{-1} + p_{n2}z^{-2}.$$

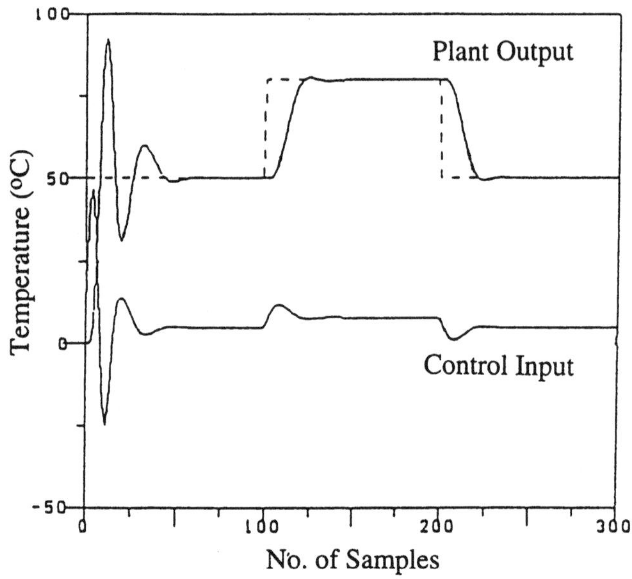

Fig. 3.4.1. Performance of the system in the simulation example using STPID controller.

The value of v is selected to obtain a somewhat stable, although oscillatory response may occur. The values of p_{d1}, p_{n1}, and p_{n2} are then selected to be -0.75, 0.18, and -0.85, respectively, to shape up the response. In applying STPID, we could expect a large initial transient which is due to the initial values of the parameter estimations. The large initial transient can be eliminated by using a conventional PID control in the initial phase of the simulation, in this case the first 20 sampling instants. Figure 3.4.1 shows the response of the STPID controller for the second order system. It can be seen that the response improves during the second and third set-points. To make some comparisons of the results, we then use a PID controller to control the system in this example.

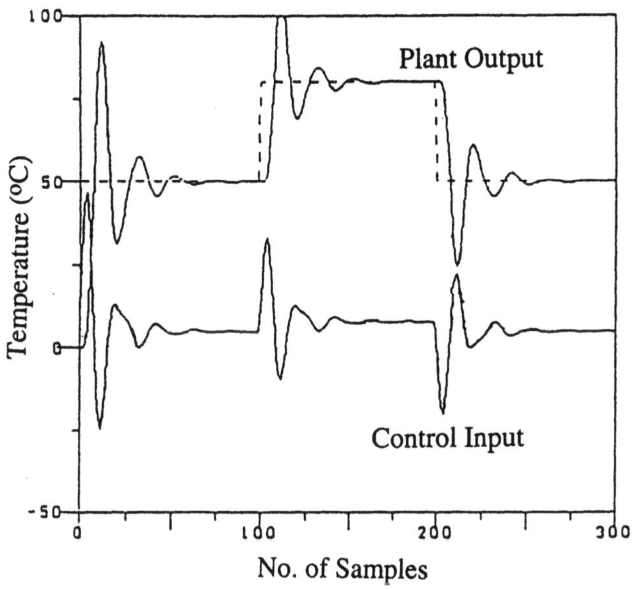

Fig. 3.4.2. Performance of the system in the simulation example using the fixed PID controller.

Figure 3.4.2 shows the behaviour of the closed loop response of the system when PID controller whose parameters are fixed is used. Although the initial transient of the response is not so large, the response does not improve in the second nor the third transients.

To make some study on the role of the tuning parameters, we first decrease the value of p_{n2} to -0.99 and we obtain a more oscillatory response as shown in Fig 3.4.3a. As expected, increasing the value of p_{n1} or p_{n2} results in an underdamped response as shown in Fig. 3.4.3b. The role of $P_d(z^{-1})$ is also studied by first decreasing p_{d1}. It can be seen from Fig 3.4.3c that the rate of change of the response is slow.

The role of p_{d1}, p_{n1}, and p_{n2} can be explained in terms of its relation to K_P, K_I, and K_D of the conventional PID controller. Referring to the identity equation in (3.3.8), we can see that for the system in this example, the value of p_{d1} is directly related to F_2 which in turn is directly related to K_D as in (3.4.7).

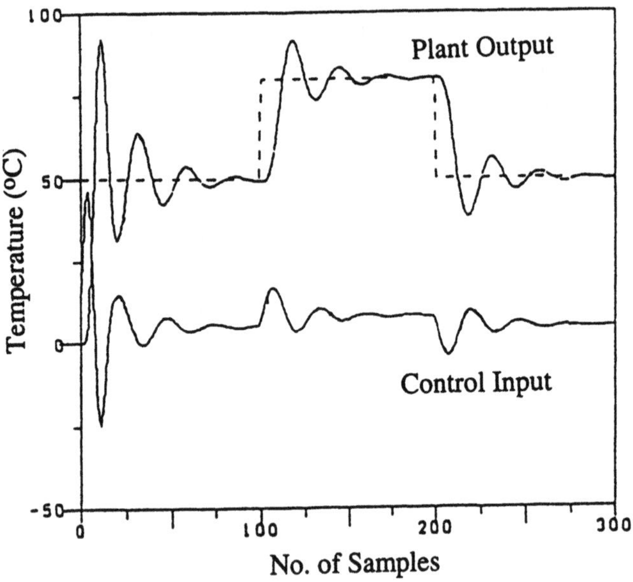

Fig. 3.4.3 (a). Performance of the STPID controller on the system with a decrease in p_{n2} value.

The effect of K_D in a conventional PID controller is to provide some anticipation of where the process is heading. It is obvious that the effect of p_{d1} is similar to K_D from the above simulation results. The tuning parameters p_{n1} and p_{n2} have some influences on the corresponding K_P and K_I referring to (3.3.8), (3.4.5), and (3.4.6).

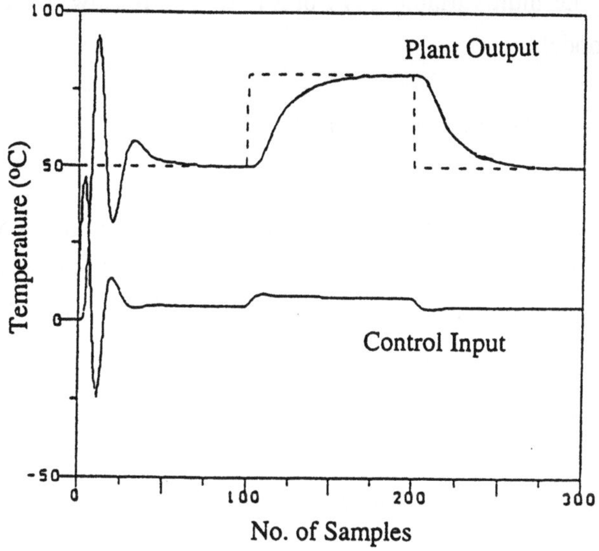

Fig. 3.4.3 (b). Performance of the STPID controller on the system with an increase in both p_{n1} and p_{n2} values.

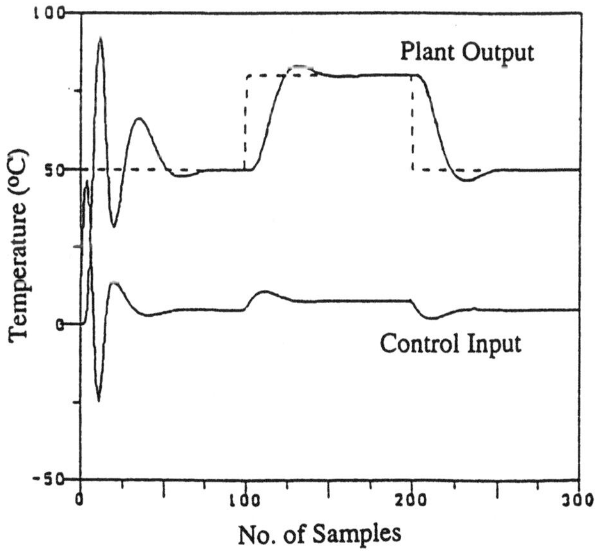

Fig. 3.4.3 (c). A simulation result of the STPID controller for the system with a decrease in p_{d1} value.

3.5 Self-Tuning PID Control: A Multivariable Approach

In deriving the multivariable STPI or STPID control law, we consider the following model

$$A(z^{-1})y(t) = B(z^{-1})u(t-d) + C(z^{-1})\xi(t) \tag{3.5.1}$$

where y(t) is an (n x 1) measured output vector, u(t) is an (n x 1) control input vector, $\xi(t)$ is an (n x 1) vector and an uncorrelated sequence of random variables with zero mean and covariance $E\{\xi(t)\xi^T(t)\} = \rho_\xi$. The (n x n) polynomial matrices $A(z^{-1})$, $B(z^{-1})$, and $C(z^{-1})$ are expressed in terms of the backward shift operator z^{-1}

$$A(z^{-1}) = I + A_1 z^{-1} + A_2 z^{-2} + \dots + A_{n_a} z^{-n_a} \tag{3.5.2}$$

$$B(z^{-1}) = B_0 + B_1 z^{-1} + B_2 z^{-2} + \dots + B_{n_b} z^{-n_b} \tag{3.5.3}$$

$$C(z^{-1}) = I + C_1 z^{-1} + C_2 z^{-2} + \dots + C_{n_c} z^{-n_c}. \tag{3.5.4}$$

There are several assumptions needed to realize this algorithm. These assumptions are summarized as follows:

A1. The noise corrupting the system is such that $C(z^{-1})$ is inverse stable (i.e., zeros of det $C(z^{-1})$ lie in the unit disc of the z-plane).

A2. $A(z^{-1})$ and $B(z^{-1})$ are relatively prime.

A3. The structural indices of the system are equal and the upper bounds for the orders of all scalar polynomials appearing in $A(z^{-1})$, $B(z^{-1})$, and $C(z^{-1})$ are known.

A4. The transport delay which is at least d time steps, associated with each transmission channel is assumed to be equal and known.

A5. $A(z^{-1})$ and $B(z^{-1})$ are stable. This means that the system is minimum phase.

In similarity to the derivation of the SISO case, the control law is derived considering the minimization of the variance of a pseudo-output $\phi(t)$ given by

$$\phi(t) = P(z^{-1})y(t) - R(z^{-1})r(t-d) + Q'(z^{-1})u(t-d) \tag{3.5.5}$$

where $R(z^{-1})$ and $Q'(z^{-1})$ are (nxn) polynomial matrices and r(t) is an (nx1) set-point vector. $P(z^{-1})$ is a rational matrix which can be represented by the following right matrix-fraction description

$$P(z^{-1}) = P_n(z^{-1})P_d(z^{-1}) \tag{3.5.6}$$

where $P_n(z^{-1})$ and $P_d(z^{-1})$ are nxn numerator and denominator matrices, respectively. $P(z^{-1})$, $R(z^{-1})$, and $Q'(z^{-1})$ are specified by the user and allow a more general closed loop to be obtained.

At this point, we let $A'(z^{-1}) = [A(z^{-1})P_d(z^{-1})]$ and we introduce (see Kailath [40], Gantmatcher [41], Maciejowski [42]) a right matrix-fraction description of the transfer function $A'(z^{-1})^{-1}C(z^{-1})$ as

$$[A'(z^{-1})]^{-1}C(z^{-1}) = \bar{C}(z^{-1})[\bar{A}'(z^{-1})]^{-1}. \tag{3.5.7}$$

We assume that $\det \bar{C}(z^{-1}) = \det C(z^{-1})$ and $\bar{A}'(0) = \bar{C}(0) = I$. We can now introduce the identity

$$P_n(z^{-1})\bar{C}(z^{-1}) = E'(z^{-1})\bar{A}'(z^{-1}) + z^{-d}F'(z^{-1}) \tag{3.5.8}$$

where $E'(z^{-1})$ and $F'(z^{-1})$ denote nxn polynomial matrices with the following form

$$E'(z^{-1}) = I + E'_1 z^{-1} + E'_2 z^{-2} \ldots + E'_{d-1} z^{-(d-1)} \tag{3.5.9}$$

and

$$F'(z^{-1}) = F'_0 + F'_1 z^{-1} + \ldots + F_{(n_a + n_{pd} - 1)} z^{-(n_a + n_{pd} - 1)} . \tag{3.5.10}$$

Here, n_{pd} is the degree of $P_d(z^{-1})$. To solve the non-commutativity of matrices, we introduce the polynomial matrices $\tilde{C}(z^{-1})$, $\tilde{E}'(z^{-1})$ and $\tilde{F}'(z^{-1})$ such that

$$E'(z^{-1})C^{-1}(z^{-1}) = \tilde{C}^{-1}(z^{-1})\tilde{E}'(z^{-1}) \tag{3.5.11}$$

$$F'(z^{-1})C^{-1}(z^{-1}) = \tilde{C}^{-1}(z^{-1})\tilde{F}'(z^{-1}) \tag{3.5.12}$$

where

$$\det C(z^{-1}) = \det \tilde{C}(z^{-1}) \text{ and } \tilde{C}(0)=I .$$

Using (3.5.5), (3.5.6), (3.5.9), and (3.5.10), we get

$$\tilde{C}(z^{-1})P_n(z^{-1}) = \tilde{E}'(z^{-1})A'(z^{-1}) + z^{-d}\tilde{F}'(z^{-1}) . \tag{3.5.13}$$

Post multiplying (3.5.13) with $P_d^{-1}(z^{-1})$ and using the model equation in (3.3.1), and (3.5.11), we have

$$P(z^{-1})y(t+d) = \tilde{C}^{-1}(z^{-1})(\tilde{F}'(z^{-1})P_d^{-1}(z^{-1})y(t) + \tilde{E}'(z^{-1})B(z^{-1})u(t)) \\ + E'(z^{-1})\xi(t+d) . \tag{3.5.14}$$

We define $\phi^*(t+d/t)$ as the optimum prediction of $\phi(t+d)$ based on the measurement up to time t and $\tilde{\phi}(t+d)$ as the prediction error. Therefore, we have

$$\phi^*(t+d/t) = \tilde{C}^{-1}(z^{-1})(\tilde{F}'(z^{-1})P_d^{-1}(z^{-1})y(t) + (\tilde{G}'(z^{-1}) + Q(z^{-1}))u(t)) - R(z^{-1})r(t) \quad (3.5.15)$$

$$\begin{aligned}\tilde{\phi}(t+d) &= \phi(t+d) - \phi^*(t+d/t) \\ &= E'(z^{-1})\xi(t+d)\end{aligned} \quad (3.5.16)$$

where $\tilde{C}(z^{-1})Q'(z^{-1})=Q(z^{-1})$ and $\tilde{E}'(z^{-1})B(z^{-1})=\tilde{G}'(z^{-1})$.

Noting that the future errors $\xi(t+1)$, ..., and $\xi(t+d)$ in (3.5.14) are uncorrelated with the present and past input and output measurements, the minimum variance control of $\phi(t+d)$ is obtained by choosing u(t) such that $\phi^*(t+d/t)$ is set to zero. We now have

$$(\tilde{G}'(z^{-1}) + Q(z^{-1}))u(t) = \tilde{C}(z^{-1})R(z^{-1})r(t) - \tilde{F}'(z^{-1})y_f(t). \quad (3.5.17)$$

Here, $y_f(t) = P_d(z^{-1})^{-1}y(t)$ and $y_f(t)$ is the filtered output. To obtain the controller parameters $\tilde{F}'(z^{-1}), \tilde{G}'(z^{-1})$, and the noise term, $\tilde{C}(z^{-1})$, directly, the quantity $\phi_y(t)$ defined by

$$\phi_y(t+d) = P(z^{-1})y(t+d) \quad (3.5.18)$$

is considered.

Using (3.5.13), an expression for $\phi_y(t+d)$ can be written as

$$\phi_y(t+d) = \tilde{C}(z^{-1})(\tilde{F}'(z^{-1})y_f(t) + \tilde{G}'(z^{-1})u(t)) + E'(z^{-1})\xi(t+d) \quad (3.5.19)$$

and

$$\phi_y^*(t+d/t) = \tilde{C}(z^{-1})^{-1}(\tilde{F}'(z^{-1})y_f(t) + \tilde{G}'(z^{-1})u(t)) \quad (3.5.20)$$

where $\phi_y^*(t+d/t)$ is the optimal d step ahead prediction of $\phi_y(t+d)$ based on the output data up to time t. Let $e(t+d) = E'(z^{-1})\xi(t+d)$. Then we have

$$e(t+d) = \phi_y(t+d) - \phi_y^*(t+d/t). \tag{3.5.21}$$

For parameter estimation purposes, we can rewrite (3.5.17) as follows:

$$\phi_y(t) = \tilde{F}'(z^{-1})y_f(t-d) + \tilde{G}'(z^{-1})u(t-d) - (\tilde{C}(z^{-1}) - I)\phi_y^*(t/d-t) + e(t) \tag{3.5.22}$$

and the data vector x(t) is defined as

$$\begin{aligned} x(t) = [&y_f^T(t),\ y_f^T(t-1), \cdots,\ u^T(t),\ u^T(t-1), \cdots, \\ &\phi_y^{*T}(t+d-1/t-1),\ \phi_y^{*T}(t+d-2/t-2), \cdots]. \end{aligned} \tag{3.5.23}$$

The corresponding parameter matrix θ(t) is given by

$$\begin{aligned} \theta(t) &= [\theta_1\ \theta_2 \ldots \theta_n]^T \\ &= [\tilde{F}'_0(t)\ \tilde{F}'_1(t) \ldots \tilde{G}_0(t)\ \tilde{G}_1(t) \ldots \tilde{C}_1(t)\ \tilde{C}_2(t) \ldots] \end{aligned} \tag{3.5.24}$$

where the column vector $\theta_i(t)$ is

$$\begin{aligned} \theta_i(t) = [&f^0_{i0}(t),\ f^0_{i1}(t), \ldots f^0_{im}(t),\ f^1_{i0}(t),\ f^1_{i1}(t), \ldots f^1_{im}(t), \ldots, \\ &g^0_{i0}(t),\ g^0_{i1}(t), \ldots g^0_{im},\ g^1_{i0}(t),\ g^1_{i1}(t), \ldots g^1_{im}(t), \ldots, \\ &c^1_{i0}(t),\ c^1_{i1}(t), \ldots c^1_{im}(t), \ldots]. \end{aligned}$$

Now $f^0_{i0}(t), g^0_{i0}(t)$, and $c^1_{i0}(t)$ denote the elements of the first row of $\tilde{F}'_0(t)$, $\tilde{G}_0(t)$ and $\tilde{C}_1(t)$, respectively.

If $\tilde{C}(z^{-1})$ is an identity matrix, then (3.5.22) contains only the quantities known at time t and the prediction error is white. Therefore, the controller parameters $\tilde{F}'(z^{-1})$ and $\tilde{G}'(z^{-1})$ can be obtained using parameter estimation scheme such as recursive least squares estimate. If $\tilde{C}(z^{-1})$ is not an identity matrix, then the parameter estimation will be biased since $\phi_y^*(t+d/t)$ is unknown. However, we can replace $\phi_y^*(t+d/t)$ by its estimated value $\hat{\phi}_y^*(t+d/t)$ which can be calculated and use the normal recursive least squares estimator to obtain the estimates of the parameters. This method is similar to using extended least squares estimator (Soderstrom and Stoica, [31], Åström and Wittenmark, [32]). The data vector is now as follows:

$$\hat{X}(t) = [y_f^T(t), y_f^T(t-1), \ldots, u^T(t), u^T(t-1), \ldots, \hat{\phi}_y^{*T}(t+d-1/t-1), \hat{\phi}_y^{*T}(t+d-2/t-2) \ldots]. \tag{3.5.25}$$

To have the structure of a multivariable PID controller in (3.5.17), it is necessary for the degree of $\tilde{F}'(z^{-1})$ to be equal to 2. Therefore,

$$\tilde{F}'(z^{-1}) = \tilde{F}'_0 + \tilde{F}'_1 z^{-1} + \tilde{F}'_2 z^{-2}. \tag{3.5.26}$$

A further requirement is to introduce an integral control action into the multivariable GMV control law. This can be done by letting

$$\tilde{G}(z^{-1}) + Q(z^{-1}) = V'(1-z^{-1}) \tag{3.5.27}$$

where

$$V' = \begin{bmatrix} V'_1 & 0 \\ 0 & V'_2 \end{bmatrix}.$$

The elimination of steady-state error is slightly different such that $\tilde{C}(z^{-1})R(z^{-1}) = H_0$ where

$$H_0 = \sum_{i=0}^{n_a+n_{pd}-1} \tilde{F}'_i Pd^{-1}(z^{-1})\Big|_{z^{-1}}. \tag{3.5.28}$$

The control law is as follows:

$$\Delta u(t) = VH_0 r(t) - V(\tilde{F}'_0 + \tilde{F}'_1 z^{-1} + \tilde{F}'_2 z^{-2}) y_f(t). \qquad (3.5.29)$$

The expression for the multivariable STPID controller parameters in terms of the multivariable PID controller parameters are given as follows:

$$K_P = V(-\tilde{F}'_1 - 2\tilde{F}'_2)\alpha \qquad (3.5.30)$$

$$K_I = V(\tilde{F}'_0 + \tilde{F}'_1 + \tilde{F}'_2)\alpha \qquad (3.5.31)$$

$$K_D = V\,\tilde{F}'_2\,\alpha \qquad (3.5.32)$$

where $\alpha = P_d^{-1}(z^{-1})\big|_{z=1}$.

For a PI or PID controller structure to be realized, the degree of $\tilde{F}'(z^{-1})$ must be 1 or 2, respectively. By reasonable assumption of the relative primeness and non-singularity of the matrices $A(z^{-1})$ and $C(z^{-1})$, we have $\tilde{E}'(z^{-1}) = E'(z^{-1})$ and $\tilde{F}'(z^{-1}) = F'(z^{-1})$ in the case of $C(z^{-1}) = I$. The degree of $\tilde{F}'(z^{-1})$ is then governed by the degree of $P_d(z^{-1})$ and also $A(z^{-1})$. It follows that if the degree of $A(z^{-1})$ is 1, then a PI or PID controller is effective if $P_d(z^{-1})$ is chosen to be of degree 1 or 2, respectively. In some applications, it is adequate to control a first order system with a PI controller. This is advantageous in that fewer parameters need to be estimated and preselected. However, some processes require derivative actions to suppress oscillations around the set-points due to slow acting controllers. The range of processes that can be controlled using this algorithm is restricted to one having a maximum order of 3. However, it can be argued that, model reduction procedure can be applied for higher order processes. Some examples of model reduction methods can be found in Isermann [43], Kokotovic *et al.* [44], and Mishra and Wilson [45]. If $C(z^{-1})$ is not an identity matrix, then $\tilde{E}'(z^{-1})$ and $\tilde{F}'(z^{-1})$ are determined by first calculating the greatest common divisor of $A'(z^{-1})$ and $C(z^{-1})$, and then using (3.5.7), (3.5.8), (3.5.11), (3.5.12), and (3.5.13). In this case, the degree of $\tilde{F}'(z^{-1})$ is not necessarily the same as $F'(z^{-1})$ (see Borisson [46]).

3.5.1 Simulation Example

The objective of this simulation example is to study the ability of the multivariable STPI or STPID controller in controlling a process under set-point changes. The process is a 2 input-2 output water bath system shown in Fig.3.5.1. The water bath system can be modeled in discrete time as follows:

$$A(z^{-1})y(t)=B(z^{-1})u(t-1)+\xi(t) \tag{3.5.33}$$

where

$$A(z^{-1}) = I+A_1z^{-1},$$

and

$$A_1 = \begin{bmatrix} -0.411 & -0.634 \\ -0.103 & -0.885 \end{bmatrix}$$

$$B(z^{-1}) = B_0 = \begin{bmatrix} 0.492 & 0.085 \\ 0.041 & 0.237 \end{bmatrix}.$$

$\xi(t)$ is uncorrelated Gaussian noise with zero mean and covariance ρ_ξ. The model is obtained by balancing the heat flow equation of the water bath system.

In applying multivariable STPI or STPID, the prefilter polynomial matrices have to be preselected on a trial and error basis. The prefilter polynomial matrices can be considered as giving phase advance in the feedback path of the control loop. This property will effect both the set-point responses and ensures a smoother disturbance rejection. It is interesting to note that the elements of $P_n(z^{-1})$ and $P_d(z^{-1})$ have a notable effect on the parameters of the controller.

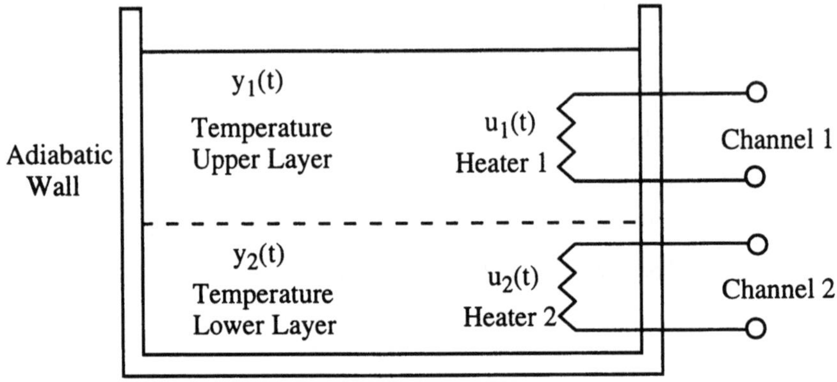

Fig. 3.5.1. A schematic diagram of a two-channel water bath system.

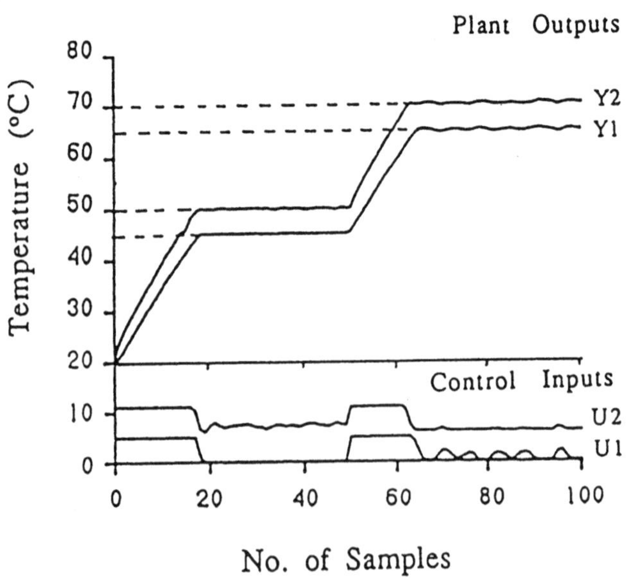

Fig. 3.5.2. Simulation result of the multivariable STPI on the water bath system.

Therefore, these polynomials can be used to shape up a better closed loop response. In a system where the interactions are low, it is adequate to choose $P_n(z^{-1})$ to be a diagonal matrix. The prefilter polynomial matrices are preselected as follows:

$$P_n(z^{-1}) = I+P_{n1}z^{-1} \text{ and } P_{n1}=\begin{bmatrix} -0.4 & 0 \\ 0 & -0.7 \end{bmatrix}$$

$$P_d(z^{-1}) = I+P_{d1}z^{-1} \text{ and } P_{d1}=\begin{bmatrix} -0.3 & 0 \\ 0 & -0.4 \end{bmatrix}.$$

V_1 and V_2 are selected to be 7 and 5, respectively. Figure 3.5.2 shows the performance of a multivariable STPI controller in controlling the temperature of the water bath under set-point changes.

3.6 Generalized Predictive Control-Some Theoretical Aspects

3.6.1 Cost Criterion

The cost criterion adopted for GPC algorithm was preceded by the cost criterion of minimum variance control (Åström and Wittenmark [6], Åström *et al.* [47]) and also GMV (Clarke and Gawthrop [18], [19]) which has been discussed in Section 3.3. The controller is obtained for a given linear input-output model, by minimizing the cost criterion of the following form:

$$J(u, t) = E[(y(t+j) - r(t+j))^2 + \lambda_1 \Delta(u(t))^2]. \qquad (3.6.1)$$

The obvious difference between the above cost criterion and that of GMV is the statement of the optimization problem which is not in terms of u(t), but in terms of $\Delta u(t)$. The term $\Delta u(t)$ has been defined in Section 3.2 which means the incremental control input of the system. At time t, the criterion is minimized with respect to $\Delta u(t)$. The use of incremental inputs in the criterion ensures the occurrence of zero static error in the case of a non-zero constant reference signal. However, the use of incremental control makes it difficult to obtain explicit bounds on u(t). Thus, u(t) may grow to be very large. To prevent such thing from happening and to remain as close as possible to the objective of maintaining the output variance minimal, the weighting λ_1 for the incremental control must be kept as small as possible while maintaining the closed loop stability. Hence, the incremental control weighting λ_1 is included simply to limit the control signal.

Another feature of GPC is the ability to control non-minimum phase and some open-loop unstable systems, and particularly systems with poorly known time-delays. The cost criterion in (3.6.1) is modified to cater for these properties as follows:

$$J1(u,t) = E\left[\sum_{j=N1}^{N2} (y(t+j)-r(t+j))^2 + \sum_{j=1}^{NU} \lambda_1(j)(\Delta(u(t+j-1))^2\right]. \tag{3.6.2}$$

with the condition that $\Delta u(t+j-1) = 0$ for $j>NU$.

NU is the control horizon, N1 is the minimum costing horizon, N2 is the maximum costing horizon, and $\lambda_1(j)$ is the control weighting sequence. The addition of the constraint on the control increments can be interpreted as effectively placing infinite weights on the control changes after some future time. This in turn not only encourage the controller to achieve its performance quickly and stabilization of non-minimum phase plant, but also significantly reduces the computational burden. Appropriate choices of the control weighting and costing horizons also enable adequate performance and asymptotic stability to be achieved for open loop stable plants.[see Bitmead *et al* . [48]).

The minimization procedure is performed on several (N2-N1+1) future output values taking into account several (NU) future incremental control actions. The success of the algorithm depends to some extend on the inclusion of the true delay in the interval between N1 and N2. Usually N1 is chosen the same as the time-delay. The minimization procedure also uses the concept of Receding Horizon Control, in which $\Delta u(t)$, $\Delta u(t+1)$,..., $\Delta u(t+NU-1)$ are produced, but only u(t) is actually applied. At time t+1, a new minimization problem is solved. The solution of the control objective of the cost criterion provides both adequate performance and asymptotic stability for a wider range of plants.

3.6.2 The Plant Model and Optimization Solution.

We first consider GPC algorithm for SISO system. In the derivation of GPC algorithm, Clarke *et al.* adopted a CARIMA model which is more appropriate to be used in process control to eliminate offset which is caused by load disturbances. Load disturbances can be modeled by drift (Brownian motion) or as random steps occurring at random instant of time. A CARIMA model with time-delay d taken as 1 can be written as follows:-

$$A(z^{-1})y(t) = B(z^{-1})u(t-1) + C(z^{-1})\frac{\xi(t)}{\Delta}. \tag{3.6.3}$$

The disturbance signal may be either a deterministic or stochastic signal but, due to the Δ operator, its mean value is assumed to be zero. In the GPC approach, $\xi(t)$ is assumed to be stochastic and the polynomial $C(z^{-1})$ is assumed as stable and equal to 1.

To solve the optimization problem of (3.6.2), we need to compute a set of j-step ahead predictions of the output y(t+j) for j= N1,...,N2, based on the information known at time t and the future values of the control increments, which will be chosen such that the criterion (3.6.2) is optimized. The j-step ahead predicted outputs can be written as follows:

$$\hat{y}(t+j/t) = G_j(z^{-1})\,\Delta\, u(t+j-1) + F_j(z^{-1})y(t) + C(z^{-1})\frac{\xi(t)}{\Delta} \tag{3.6.4}$$

where $G_j(z^{-1})=E_j(z^{-1})B(z^{-1})$, E_j and F_j are obtained by solving the Diophantine equation recursively as follows:

$$1 = E_j(z^{-1})\,A\Delta + z^{-j}F_j(z^{-1}) \tag{3.6.5}$$

$$1=\tilde{R}_j(z^{-1})\,A\Delta+z^{-(j+1)}\tilde{S}_j(z^{-1}) \tag{3.6.6}$$

where $\tilde{R}_j(z^{-1})=E_{j+1}(z^{-1})$ and $\tilde{S}_j(z^{-1})=F_{j+1}(z^{-1})$, $j=1,2,\ldots N$.

Now subtracting (3.6.5) from (3.6.6), we have

$$0=(\tilde{R}_j(z^{-1})-E_j(z^{-1}))A(z^{-1})\Delta+z^{-j}(z^{-1}\tilde{S}_j(z^{-1})-F_j(z^{-1})). \tag{3.6.7}$$

Now

$$\begin{aligned}\tilde{R}_j(z^{-1})-E_j(z^{-1}) &= (1+\tilde{r}_1z^{-1}+\tilde{r}_2z^{-2}+\ldots\ \tilde{r}_jz^{-j})-\\ &\qquad (1+e_1z^{-1}+e_2z^{-2}+\ldots\ e_{j-1}z^{-(j-1)})\end{aligned}$$

$$=(\tilde{r}_1-e_1)z^{-1}+(\tilde{r}_2-e_2)z^{-2}+\ \ldots+(\tilde{r}_{j-1}-e_{j-1})z^{-(j-1)}+\tilde{r}_jz^{-j}$$

$$=\bar{R}_j(z^{-1})+\tilde{r}_jz^{-j}. \tag{3.6.8}$$

Letting $\mathbf{A}(z^{-1})=A(z^{-1})\Delta$, we now have

$$\mathbf{A}\bar{R}=(1+\mathbf{a}_1z^{-1}+\mathbf{a}_2z^{-2}+\ldots+\mathbf{a}_{na+1}z^{-(na+1)})(\bar{r}_1z^{-1}+\bar{r}_1z^{-2}+\ldots+\bar{r}_{j-1}z^{-(j-1)})$$

$$\begin{aligned}&=(\bar{r}_1z^{-1}+(\bar{r}_2+\mathbf{a}_1\bar{r}_1)z^{-2}+\ldots+(\bar{r}_{j-1}+\mathbf{a}_1\bar{r}_{j-2}+\ldots)z^{-(j-1)}\\ &\quad+\ldots+\mathbf{a}_{na+1}(\bar{r}_{j-1})z^{-(j+na)}).\end{aligned} \tag{3.6.9}$$

Now from (3.6.7) we have

$$\mathbf{A}(z^{-1})\bar{R}(z^{-1})+z^{-j}(z^{-1}\tilde{S}(z^{-1})-F(z^{-1})+\mathbf{A}(z^{-1})\tilde{r}_j)=0. \tag{3.6.10}$$

Since $z^{-j}(z^{-1}\tilde{S}(z^{-1}) - F(z^{-1}) + \mathbf{A}(z^{-1})\tilde{r}_j)$ contains the terms z^{-j}, $z^{-(j+1)}, \ldots$, therefore, the first (j-1)-th term of $\mathbf{A}(z^{-j})\overline{R}(z^{-1})$ is zero and hence, $\overline{R}(z^{-1}) = 0$. Now we have

$$z^{-j}(z^{-1}\tilde{S}(z^{-1}) - F(z^{-1}) + \mathbf{A}(z^{-1})\tilde{r}_j) = 0$$

and

$$\tilde{S}(z^{-1}) = z(f_0 + f_1 z^{-1} + .. + \tilde{r}_j - \mathbf{a}_1 \tilde{r}_j z^{-1} - .. \mathbf{a}_{na+1} \tilde{r}_j z^{-(na+1)}).$$

Equating the z term, we have

$$f_0 = \tilde{r}_j \text{ and } \tilde{s}_j = f_{i+1} - \mathbf{a}_{i+1}\tilde{r}_j$$

where $i = 0$ to deg of $\tilde{S}(z^{-1})$. (3.6.11)

The predicted output can be decomposed into two components, i.e., $\mathbf{f}(t+j)$ which is known at time t and $\Gamma\Delta u(t+j-1)$ is the component due to the control input. Therefore,

$$Y = \Gamma U + \mathbf{F} \tag{3.6.12}$$

where $Y = [y(t+1)\ y(t+2)\ \ldots y(t+N)]^T$

$U = [\Delta\, u(t)\,,\ \Delta\, u(t+1) \ldots . \Delta u(t+N-1)]^T$

$\mathbf{F} = [\mathbf{f}(t+1), \mathbf{f}(t+2) \ldots \mathbf{f}(t+N)]^T$.

Γ is a lower triangular matrix of dimension NxN and its element are the plant's step response,

$$\Gamma = \begin{bmatrix} \Gamma_0 & 0\ldots\ldots\ldots & 0 \\ \Gamma_1 & & \\ \cdot & \Gamma_0 & \cdot \\ \cdot & & \cdot \\ \cdot & & \cdot \\ \Gamma_{N2-1} & \ldots\ldots\ldots & \Gamma_{N2-NU} \end{bmatrix}.$$

A distinct feature of the GPC algorithm is that choices of the design parameters N1, N2, and λ_1 are flexible and do not affect the stability of the controlled system. However, reduction in computation time will result if appropriate values of N1 and N2 are chosen.

The quadratic minimization problem of (3.6.2) now becomes

$$J1 = E[(Y-\mathbf{r})^T(Y-\mathbf{r}) + \lambda_1 U^T U] \tag{3.6.13}$$

where $\mathbf{r}$ is the vector for the pre-specified set-points. The solution of the future incremental control vector $\Delta u(t)$ is obtained by partial differentiating the cost function with respect to U. We now have

$$U = (\Gamma^T\Gamma + \lambda_1^{-1}I)^{-1}\Gamma^T(\mathbf{r}-\mathbf{f}). \tag{3.6.14}$$

Since only the first control is required, then

$$u(t) = u(t-1) + \overline{\Gamma}^T(\mathbf{r}-\mathbf{f}) \tag{3.6.15}$$

and $\overline{\Gamma}^T$ is the first row of $(\Gamma^T\Gamma + \lambda_1^{-1}I)^{-1}\Gamma^T$.

To make the algorithm self-tuning, the parameters of the plant are estimated recursively by means of recursive least squares estimates discussed in Section 3.2. The data and the parameter vectors of the plant model, respectively are as follows:

$$x^T=[\Delta u(t-1), \Delta u(t-2),..., \Delta y(t-1), \Delta y(t-2),...] \quad (3.6.16)$$

$$\theta^T = [b_0, b_1,..., a_1, a_2,...]. \quad (3.6.17)$$

3.7 Fuzzy Logic Control

In recent years, the term "fuzzy" has dominated a wide variety of domestic products and industrial systems. Much of these developments is mainly due to a better understanding of the concept of fuzzy logic, its simplicity of implementation, and its feasibility in hardware development. Originally, when fuzzy logic was first conceived [30], [49],[50], it was expected that most of its applications would be in the realm of knowledge-based systems in which the resident information is both imprecise and uncertain. Contrary to this expectation, however, most of the successful applications of fuzzy logic until today has been related to systems control in which there is imprecision but no uncertainty.

The idea of applying fuzzy logic to control systems was first conceived by Mamdani and his colleagues [51], [52], [53], and [54]. Based on Zadeh's fuzzy set theory [30], [49], and the simple conventional PID controller, Assilian and Mamdani [51] developed what is now referred to as the basic fuzzy logic controller, which is used to regulate the outputs of a process around a given set-point using a digital computer. Due to the limitations of memory space and speed of the sequential Von Neumann computers, fuzzy logic controllers were not very popular in the beginning and much of their applications centered around slow varying processes. Some of the earlier applications of fuzzy logic to process control can be found in Mamdani [55] and Tong [56].

There are a number of advantages of applying fuzzy logic to the control of industrial processes over traditional controllers. Perhaps one of the main advantages of applying fuzzy logic control is that a controller can be developed along linguistic lines which has close associations with the field of artificial intelligence. As described by Wiener in his book [57], artificial intelligence (or neural science) and control science were once regarded as a common subject. One of the aims of artificial intelligence is to replace human beings carrying out precise tasks by machines and hence, the link between these two sciences is strong. The fuzzy controller consists of a set of linguistic conditional statements or rules (referred to as fuzzy association matrix rules or FAM rules) which define the individual control situations. These linguistic conditional statements can be easily developed from common sense or from engineering judgment of the process to be controlled.

Many industrial processes are difficult to be controlled accurately and it has been claimed that fuzzy logic control can deal successfully with such processes which are usually multi-variable, inherently nonlinear, and time-varying in nature. In addition, the fuzzy logic controller can also deal with ill-defined systems of unknown dynamics as they do not require a priori mathematical model of the plant for implementation, as required by many traditional adaptive controllers. Another advantage is that fuzzy logic controllers can now be feasibly implemented in digital or analog VLSI circuitry (Togai and Watanabe [58], Yamakawa [59], Kuo *et al.* [60], etc.) where sampled information can be encoded in a parallel-distributed framework.

3.7.1 Brief Overview of Fuzzy Set and Fuzzy System Theory

Fuzzy-set theory is an active research area and some of its aspects have bee discussed in the following literature (Dubois and Prade [61], Klir and Folg([62], Zimmermann [63], Kosko [64], and Pedrycz [65]). In this section a bri(overview of fuzzy-set theory is described from the viewpoint of between-cub theory of Kosko [64].

As stated by Kosko [64], a fuzzy set can be defined as a point in a cube whereas a fuzzy system can be defined as a mapping between these cubes. S is a fuzzy system that maps fuzzy sets to fuzzy sets. Thus, a fuzzy system transforms fuzzy sets from one hyper cube I^n to another hyper cube I^p as follows:

$$S: I^n \rightarrow I^p \tag{3.7.1}$$

where I^n which is the n-dimensional unit hyper cube contains all the fuzzy subsets of the domain space or input universe of discourse and I^p contains all the fuzzy subsets of the range space or output universe of discourse. In general, a fuzzy system S maps families of fuzzy sets to families of fuzzy sets, thus,

$$S: I^{n1} \times \ldots \times I^{nr} \rightarrow I^{p1} \times \ldots \times I^{p1}. \tag{3.7.2}$$

These fuzzy systems behave as associative memories where they map close inputs to close outputs in which they are referred to as fuzzy associative memories (FAM). The simplest FAM encodes the FAM rule or association (A_i, B_i) which associates the p-dimensional fuzzy set B_i with the n-dimensional fuzzy set A_i. These minimal FAMs essentially map one ball in I^n to one ball in I^p.

Three common associations in fuzzy set theory [30] are 'complement', 'union', and 'intersection' of fuzzy sets. The complement of a fuzzy set A is denoted by A' and is defined by

$$m_{A'}(x) = 1 - m_A(x), \quad x \in X \tag{3.7.3}$$

where X is the real line R^1 or universe of discourse and $m_A(x)$ is the membership function of the fuzzy set A between 0 and 1. The union of two fuzzy sets A and B with respective membership functions $m_A(x)$ and $m_B(x)$ is a fuzzy set C, written as $C = A \cup B$, whose membership function is related to those of A and B by

$$m_C(x) = \mathrm{Max}[m_A(x), m_B(x)], \quad x \in X \tag{3.7.4}$$

or in abbreviated form

$$m_C = m_A \vee m_B. \tag{3.7.5}$$

The intersection of two fuzzy sets A and B with respective membership functions $m_A(x)$ and $m_B(x)$ is a fuzzy set C, written as $C = A \cap B$, whose membership function is related to A and B by

$$m_C(x) = \text{Min}[m_A(x), m_B(x)], \quad x \in X \tag{3.7.6}$$

or in abbreviated form

$$m_C = m_A \wedge m_B. \tag{3.7.7}$$

In simple words, the union of A and B is the smallest fuzzy set containing both A and B, whereas the intersection of A and B is the largest fuzzy set which is contained in both A and B.

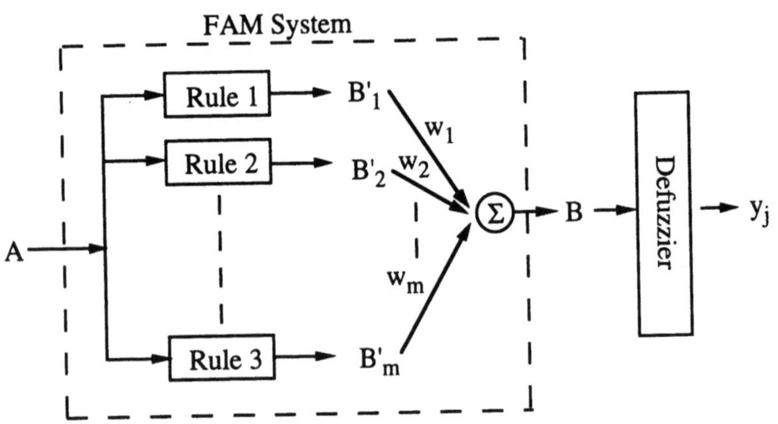

Fig. 3.7.1. A FAM system architecture. The FAM associations or rules (1 to m) are fired in parallel to produce the output fuzzy set B. B is "defuzzified" to a numerical value y_j.

In general, a FAM system F: $I^n \rightarrow I^p$ encodes and processes in parallel a FAM bank of m FAM rules $(A_1, B_1), \ldots, (A_m, B_m)$ as shown in Fig. 3.7.1. Each input A to the FAM system activates each stored FAM rule to different degree. The minimal FAM that stores the ith=*i*th or *i*th association or rule (A_i,

B_i) maps input A to B_i', which is a partially activated version of B_i. The more A resembles A_i, the more B_i' resembles B_i. The corresponding output fuzzy set B combines these partially activated fuzzy sets B_i' ... B_m' where it equals a weighted average of the partially activated sets:

$$B = \sum_{i=1}^{m} w_i B_i' \, . \tag{3.7.8}$$

The term w_i reflects the frequency or strength of the fuzzy association (A_i, B_i). In practice, the output waveform B can be "defuzzified" to a single numerical value y_j in Y by computing the fuzzy centroid of B with respect to the output universe of discourse Y.

3.7.2 Basic Concept of Fuzzy Logic Controller

A fuzzy logic control system is shown as in Fig. 3.7.2. It consists of a set of FAM rules which define individual control situations. In many applications, it is sufficient to relate the error and the rate of change of error in the process to apply the correct amount of change in the control input to satisfactorily control the system. In this respect simple linguistic rules can be formulated based on observation or a simple study about the process. An example of a linguistic natural language rule that an expert might use to describe a control system action is as follows:

If error is positive large and rate of change of error is negative small, then change in the process input is positive large.

These three variables, error (E), rate of change of error (C), and change in control input (U) are defined by fixed universes of discourse which define the range of measurements they can take in a particular control system. In some applications (Pacini and Kosko [66] and Kong and Kosko [67]), another variable such as the past control input value may be added as the antecedents in the FAM rule to improve the system performance, however, at the expense of added complexity in the development of the fuzzy control system.

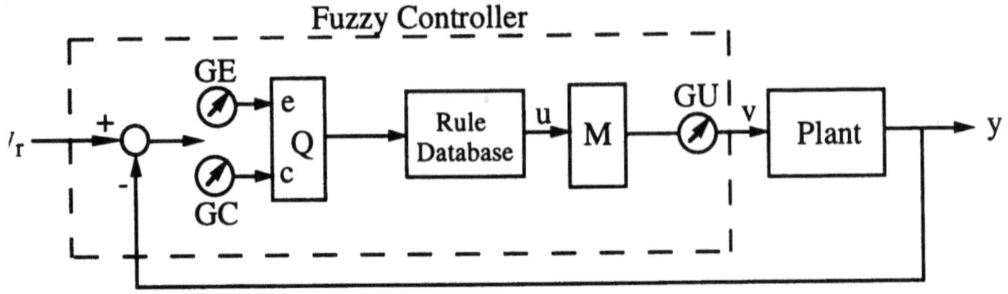

Fig. 3.7.2. A fuzzy logic control system

Each of the fuzzy variables can be quantized into fuzzy subsets according to the complexity of the process. Linguistic terms such as positive large, negative small, zero, etc. in which an expert might use to describe the control system actions are used to label these fuzzy subsets which in reality represent numerical values. Each of these fuzzy subsets contains elements with degrees of membership. A fuzzy membership function assigns a real number generally between 0 and 1 to every element in its universe of discourse which indicates the degree to which the object or data belongs to that fuzzy set.

As an example, a FAM rule k can be abbreviated to a statement of the form

$$\text{If E is } E_k \text{ and C is } C_k\text{, then U is } U_k. \tag{3.7.9}$$

These fuzzy subsets E_k, C_k, and U_k are defined by a set of ordered pairs as follows:

$$\begin{aligned} E_k &= \{(e, m_{e_k}(e))\} \subset E \\ C_k &= \{(c, m_{c_k}(c))\} \subset C \\ U_k &= \{(u, m_{u_k}(u))\} \subset U \end{aligned} \tag{3.7.10}$$

where e, c, and u are elements of the discrete universes and m_{e_k}, m_{c_k} and m_{u_k} are the corresponding membership values which give the degree to which the element is a member of the fuzzy subset.

Fuzzy membership functions can have different shapes, such as monotonic, triangular, trapezoidal, and Gaussian, according to the designer's preference or experience. Figure 3.7.3 shows four common examples of membership function shapes or splines. Selection of the type of membership functions affects the type of reasoning to be performed (see Langari and Berenji [68]). In practice, trapezoidal or triangular shapes are widely used which simplify computation.

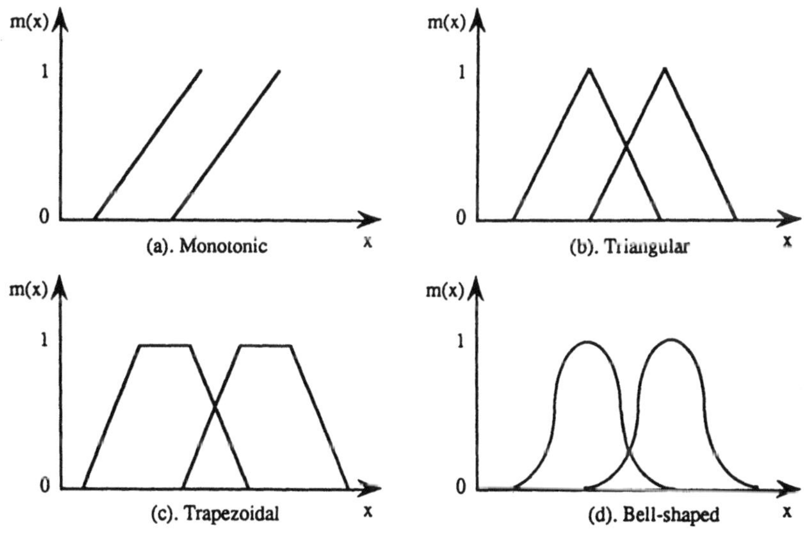

Fig. 3.7.3. Four common examples of fuzzy membership function splines.

The antecedents of all the FAM rules which describe the process control action are activated in parallel (as shown in Fig. 3.7.1) where the pulse nature of the inputs picks off single fit values of the fuzzy variables. The antecedent fit values can be combined with either minimum or maximum depending on whether the antecedent fuzzy sets are combined with the conjunctive AND or disjunctive OR. The example as described above combined the antecedents with the conjunctive AND which activates the consequent fuzzy set U_k to degree w_k, therefore, using Zadeh's fuzzy intersection association of (3.7.6), we have

$$w_k = m_{e_k}(e) \wedge m_{c_k}(c). \qquad (3.7.11)$$

The output fuzzy set's shape depends on the FAM rule encoding scheme used. Two types of encoding techniques widely used are the correlation-minimum encoding and the correlation-product encoding. With correlation-minimum encoding, the consequent fuzzy set U_k in the library of the output fuzzy set values is clipped to degree w_k with pointwise minimum such that

$$m_{O_k}(y) = w_k \wedge m_{U_k}(y) \qquad (3.7.12)$$

and with correlation-product encoding, we multiply m_{U_k} by w_k

$$m_{O_k}(y) = w_k \, m_{U_k}(y) \qquad (3.7.13)$$

where $m_{O_k}(y)$ is the degree of membership of the activated output fuzzy set at interval y in the output universe of discourse. Figure 3.7.4 shows an example of how the output is affected when (a) correlation-minimum and (b) correlation-product encoding techniques are used.

In practice, $m_{O_k}(y)$ is not unique, the system then combines all of these functions as their union to yield the actual output fuzzy membership function, $m_O(y)$, following Zadeh's fuzzy union association of (3.7.4) such that

$$m_O(y) = m_{O_1}(y) \vee m_{O_2}(y) \vee \;\ldots\ldots\; \vee m_{O_k}(y). \qquad (3.7.14)$$

Note that k in this case is not the number of rules, it is however, the quantization of the output fuzzy set into its fuzzy subsets.

The defuzzified output, i.e., the change in control output u, equals the fuzzy centroid as follows:

$$u = \frac{\int y m_O(y) dy}{\int m_O(y) dy} \tag{3.7.15}$$

where the limits of the integration correspond to the entire universe of discourse Y.

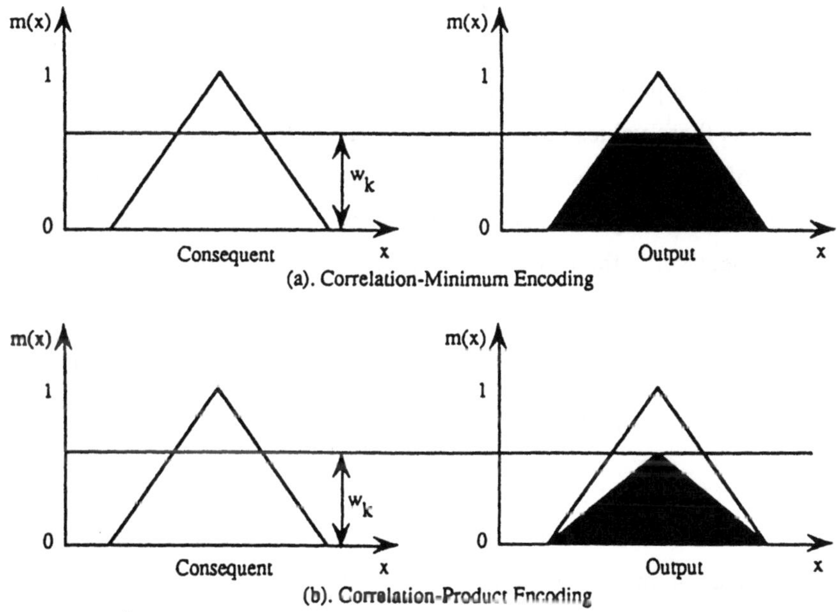

Fig. 3.7.4. Example of two FAM-inference procedures showing how the output is affected: (a) correlation-minimum and (b) correlation-product encoding.

Alternatively, in order to reduce computation, Y can be discretized to p values at regular interval, Δy, such that, $Y = \{y_1, y_2, ...y_p\}$ which gives the discrete fuzzy centroid

$$u = \frac{\sum_{j=1}^{p} y_j m_O(y_j)}{\sum_{j=1}^{p} m_O(y_j)}. \tag{3.7.16}$$

A fuzzy logic controller has a number of parameters which must be chosen by the designer in prior. In this case, the parameters are three scale factors, GE, GC, and GU for the process error, change in error, and the controller's output, respectively. The values of error and change in error used by the controller are elements of the corresponding universes of discourse which can be represented in time by e(t) and c(t) where t is the sample number. As these values have to be quantized to the closest element of the universe of discourse, they first have to be scaled by multiplying them by a suitable value. If the process output and the set-point at the time t are x(t) and $y_r(t)$, respectively, then e(t) and c(t) are obtained from the following equations:

$$e(kT) = Q[\{y_r(t) - x(t)\}(-GE)] \tag{3.7.17}$$

$$c(kT) = Q[\{x(t) - x(t-1)\}(-GC)]. \tag{3.7.18}$$

Similarly, at the output of the fuzzy controller the result of the fuzzy centroid defuzzification must be multiplied by a certain value, GU, to calculate the actual change in the process input. Thus, if v(t+1) is the new process input, it is given by

$$v(t+1) = v(t) + GUu(t). \tag{3.7.19}$$

The selection of these scale factors is akin to the selection of the PID controller parameters and the user defined polynomials of some adaptive controllers. Systematic approaches on the selection of these scale factors have been proposed by some researchers (Procyk and Mamdani [69], Daley and Gill [70], and Linkens and Abbod [71].

3.8 Conclusions

This chapter discusses several well known traditional control approaches which are the conventional PID control, self-tuning control, self-tuning PID control, generalized predictive control, and fuzzy logic control. It is hoped that the reader can understand some of the underlying principles of these traditional control approaches to be able to make comparison, and to decipher their advantages and disadvantages with some neuro-control approaches that will be discussed in the following chapter.

3.9 References

[1] Ziegler, J.G. and N.B. Nichols, "Optimum settings for automatic controllers", *Trans. of ASME*, Vol. 15, pp. 827-834, 1942.

[2] Takahashi, Y. and C.S. Chan, "Parametereinstellung bei linearen DDC-algorithmen", *Regelungstechnik und Prozess-Datenverarbeitung*, Vol.19, pp. 237-244, 1971.

[3] Nishikawa, Y., N. Sanomiya, T. Ohta, and H. Tanaka, "A method for auto-tuning of PID control parameters", *Automatica*, Vol. 20, pp. 321-332, 1984.

[4] Åström, K.J., and T. Hagglung, "Automatic tuning of simple regulators with specifications on phase and amplitude margins", *Automatica*, Vol. 20, pp. 645-651, 1984.

[5] Kalman, R.E., "Design of self-optimizing control system", *Trans. of ASME*, Vol. 80, pp. 468-478, 1958.

[6] Åström, K.J. and B. Wittenmark, "On self-tuning regulators", *Automatica*, Vol. 9, pp. 185-199, 1973.

[7] Wittenmark, B., "Self-tuning PID controllers based on pole placement", *Lund Inst. of Technical Re*port, TFRT-7179, 1979.

[8] Savelli, J.C., K. Warwick, and J.H. Wescott, "Implementation of an adaptive PID self-tuning controller", *Proc. of Melcon*, Athens, Vol. 2, 1983.

[9] Gawthrop, P.J., "Self-tuning PID controllers: Algorithm and implementation", *IEEE Trans. on Automatic Control,* Vol. 31, pp. 201-209, 1986.

[10] Banyasz, C. and L. Keviczky, "Direct methods for self-tuning PID regulators", *Proc. of the 6th IFAC Symposium on Identification & System Parameter Estimation* (Bekey, G.A. and G.N. Saridis, Eds.), Washington, D.C., pp. 1395-1400, 1982.

[11] Ortega, R. and R. Kelly,"PID self-tuners: Some theoretical and practical aspects", *IEEE Trans. on Industrial Electronics,* Vol. IE-31, pp. 312-332, 1984.

[12] Radke, F. and R. Isermann,"A parameter-adaptive PID-controller with stepwise parameter optimization", *Automatica*, Vol. 23, pp. 449-457, 1987.

[13] Jones, A.H. and B. Porter, "Design of adaptive digital set-point tracking PID controllers incorporting recursive step-response matrix identifiers for multivariable plants", *IEEE Trans. on Automatic Control,* Vol. AC-32, pp. 459-463, 1987.

[14] Proudfoot, C.G., P.J. Gawthrop, and O.L.R. Jacobs, "Self-tuning PI control of a pH neutralisation process", *Proc. of IEE,* Pt-D, Vol. 130, pp. 267-272, 1983.

[15] Yamamoto, T., S. Omatu, and T. Hotta, "A construction of self-tuning PID control algorithm and its application", *Proc. of SICE*, Japan, pp. 1143-1146, 1983.

[16] Rad, A.B. and P.J. Gawthrop, "Explicit PID self-tuning control for systems with unknown time-delay", *Proc. of IFAC Int. Symposium on Intelligent Tuning and Adaptive Control,* Singapore, 1991.

[17] Cameron, F. and D.E. Seborg, "A Self-tuning controller with a PID structure", *Int. Journal of Control*, Vol. 30, pp. 401-417, 1983.

[18] Clarke, D.W. and P.J. Gawthrop,"Self-tuning controller", *Proc. of IEE,* Pt-D, Vol. 122, pp. 929-934, 1975.

[19] Clarke, D.W. and P.J. Gawthrop, "Self-tuning control", *Proc. of IEE,* Pt-D, Vol. 126, pp. 633-640, 1979.

[20] Johansson, R., "Multivariable adaptive control", *Dept. of Automatic Control, Lund Inst. of Tech.*, Report LUFTD2/TRFT-1024/1-207, 1983.

[21] Richalet, J.A., J.L.T. Rault, and J. Papon,"Model predictive heuristic control: applications to industrial processes", *Automatica*, Vol. 14, pp. 413-428, 1976.

[22] Cutler, C.R. and B.C. Ramaker,"Dynamic matrix control - A computer control algorithm", *JACC*, Paper WP5-B, San Francisco, 1986.

[23] De Keyser, R.M.C. and A.V. Cauwenberghe,"A self-tuning multi step predictor application", *Proc. 6th IFAC Symposium on Identification and System Parameter Estimation*, Washington, DC, pp. 1558-1563, 1979.

[24] Ydstie, B., "Extended horizon adaptive control", *Proc. of 9th IFAC World Congress*, Budapest, Hungary, 1984.

[25] Peterka, V., "Predictor-based self-tuning control", *Automatica*, Vol. 20, pp. 39-50, 1984.

[26] Clarke, D.W., C. Mohtadi, and P.S. Tuffs, "Generalized predictive control - Part I: The basic algorithm", *Automatica*, Vol. 23, pp. 137-148, 1987.

[27] Clarke, D.W., C. Mohtadi, and P.S. Tuffs, "Generalized predictive control - Part II: Extensions and interpretations", *Automatica*, Vol. 23, pp. 149-160, 1987b.

[28] Clarke, D.W. and C. Mohtadi, "Properties of generalized predictive control", *Automatica*, Vol. 25, pp. 859-875, 1989.

[29] De Keyser, R.M.C., PH.G.A. Van De Velde, and F.A.G. Dumartier, "A comparative study of self-adaptive long range predictive control methods", *Automatica*, Vol. 24, pp. 49-163, 1988.

[30] Zadeh, L.A., "Fuzzy sets", *Information and Control*, Vol. 8, pp.338-352, 1965a.

[31] Söderström, T. and P. Stöica, "*System Identification* ", Prentice-Hall, London, 1989.

[32] Åström, K.J. and B. Wittenmark, *"Adaptive Control."*, Addison-Wesley, New York, 1989.

[33] Bierman, G.J., *"Factorization Methods for Discrete Sequential Estimation "*, Academic Press, New York, 1977.

[34] Gawthrop, P.J., "Some interpretations of the self-tuning controller", *Proc. of IEE,* Pt-D, Vol. 124, pp. 889-894, 1977.

[35] Clarke, D.W., "Introduction to self-tuning controllers", *In Harris and Billings (Eds.)*, Peter Pegrinus, (UK), pp. 36-71, 1981.

[36] Clarke, D.W., A.J.F. Hodgson, and P.S. Tuffs, "Offset problem and K-incremental predictors in self-tuning control", *Proc. of IEE,* Pt-D, Vol. 130, pp. 217-225, 1983.

[37] Tuff, P.S. and D.W. Clarke,"Self tuning control of offset-A unified approach". *Proc. of IEE,* Pt-D , Vol. 132, pp. 100-110, 1985.

[38] Clarke, D.W., "Self-tuning control of non minimum-phase systems" *Automatica*, Vol. 20, pp. 501-517, 1984.

[39] Clarke, D.W., P.P. Kanjilal, and C. Mohtadi, "A generalised LQG approach to self-tuning control, part 1: Aspects of design" *Int. Journal of Control,* Vol. 45, pp. 1509-1523, 1985.

[40] Kailath, T., *"Linear Systems"*, Prentice-Hall, Englewood Cliffs, New Jersey, 1980.

[41] Gantmacher, F.R., *"The Theory of Matrices"*, Chelsea Publishing Company, New York, 1977.

[42] Maciejowski, J.M., "*Multivariable Feedback Design*", Addison-Wesley, London, 1989.

[43] Isermann, R., *"Digital Control Systems: Vol. I and II"*, Springer-Verlag, Berlin, 1989.

[44] Kokotovic, P.V., R.E.Jr. O'Malley, and P. Sannuti, "Singular perturbations and order reduction in control theory - An overview", *Automatica,* Vol 12, pp. 123-132, 1976.

[45] Mishra, R.N. and D.A. Wilson, "A New Algorithm for Optimal Reduction of Multivariable Systems", *Int. Journal of Control,* Vol. 31, pp. 443-466, 1980.

[46] Borrison, U., "Self tuning for a class of multivariable systems", *Automatica*, Vol. 15, pp. 209-215, 1979.

[47] Åström, K.J., U. Borrison, L. Ljung, and B. Wittenmark, "Theory and applications of self-tuning regulators", *Automatica,* Vol. 13, pp 457-476, 1977.

[48] Bitmead, R.R., M. Gevers, and V. Wertz, *"Adaptive Optimal Control",* Prentice-Hall, Australia Pty Ltd., 1990.

[49] Zadeh, L.A., "Fuzzy sets and systems", *Proc. of Symp. on System Theory,* Polytechnic Institute of Brooklyn, pp.29-37, 1965.

[50] Zadeh, L.A., "Outline to a new approach to the analysis complex systems and decision processes", *IEEE Trans. on Systems, Man, and Cybernetics.* Vol. 3, pp. 28-44, 1973.

[51] Assilian, S. and E.H. Mamdani, "An experiment in linguistic synthesis with a fuzzy logic controller", *Int. Journal of Man-Machine Studies*, Vol. 7, pp.1-13, 1974.

[52] Mamdani, E.H., "Applications of fuzzy algorithms for a simple dynamic plant", *Proc. of IEE.,* Vol. 121, pp. 1585-1588, 1974.

[53] Mamdani, E.H , "Advances in the linguistic synthesis of fuzzy controllers", *IEEE Trans. on Computer*, Vol. C-26, pp.1182-1191, 1977.

[54] King, P.J. and E.H. Mamdani, "The application of fuzzy control systems to industrial processes", *Automatica*, Vol. 13, pp. 235-242, 1977.

[55] Mamdani, E.H., "Application of fuzzy set theory to control systems: a survey", *Fuzzy Automata and Decision Proceses* (Gupta, M.M., G.N. Sardis, and B.R. Gaines, Eds.), North-Holland, 1977.

[56] Tong, R. M. "A control engineering review of fuzzy systems", *Automatica*, Vol. 13, pp. 559-569, 1977.

[57] Wiener, N., *"Cybernetics: on control and communication in the animal and the machine"*, MIT Press, Cambridge, MA, 1948.

[58] Togai, M. and H. Watanabe, "Expert system on a chip: an engine for real-time approximate reasoning", *IEEE Expert*, Vol. 1, pp. 55-62, 1986.

[59] Yamakawa, T., "Fuzzy microprocessors - rule chip and defuzzification chip" *Proc. of Int. Workshop on Fuzzy Systems Applications*, Iizuka-88, Japan, pp. 51-52, 1988.

[60] Kuo, Y.H., C.I. Kao, and J.J. Chen, "A fuzzy neural network model and its hardware implementation", *IEEE Trans. on Fuzzy Systems*, Vol. 1, pp. 171-183, 1993.

[61] Dubois, D. and H. Prade, *"Fuzzy Sets and Systems: Theory and Applications"*, Academic Press, New York, 1980.

[62] Klir, G. and T. Folger, *"Fuzzy Sets: Uncertainty and Information"*, Prentice-Hall, Engelwood Cliffs, New Jersey, 1988.

[63] Zimmermann, H., *"Fuzzy Set Theory and its Applications"*, 2nd. Ed., Kluwer, Boston, 1990.

[64] Kosko, B., *"Neural Networks and Fuzzy Systems: A Dynamical Systems Approach to Machine Intelligence"*, Prentice-Hall, Englewood Cliffs, New Jersey, 1991.

[65] Pedrycz, W., *"Fuzzy Control and Fuzzy Systems"*, 2nd Ed., Wiley, Somerset, 1993.

[66] Pacini, P.J. and B. Kosko "Comparison of fuzzy and Kalman-filter target-tracking control systems", in *Neural Networks and Fuzzy Systems: A Dynamical Systems Approach to Machine Intelligence*, Prentice-Hall, Englewood Cliffs, New Jersey, pp. 379-406, 1991.

[67] Kong, S.G. and B. Kosko, "Adaptive fuzzy systems for backing up a truck-and-trailer", *IEEE Trans. on Neural Networks*, Vol. 3, pp. 211-223, 1992.

[68] Langari, R. and H.R. Berenji "Fuzzy logic in control engineering", in *Handbook of Intelligent Control: Neural, Fuzzy, and Adaptive Approaches*. (White, D.A. and D.A. Sofge, Eds.), Van Nostrand Reinhold, New York, 1992.

[69] Procyk, T.J. and E.H. Mamdani, "A linguistic self-organizing process controller", *Automatica*, Vol. 15, pp. 15-30, 1979.

[70] Daley, S. and K.F. Gill, "A design study of a self-organizing fuzzy logic controller", *Proc. of I. Mech . E*, Vol. 200, (CI), pp.59-69, 1986.

[71] Linkens, D.A. and M.F. Abbod,"Self-organizing fuzzy logic control and the selection of its scaling factors", *Trans. on Instrumentation, Measurement and Control,* Vol. 14, pp. 114-125, 1992.

CHAPTER 4

NEURO-CONTROL TECHNIQUES

4.1 Introduction

In many real-world applications, there are many nonlinearities, unmodeled dynamics, unmeasurable noise, multiloop, etc., which pose problems to engineers in trying to implement control strategies. During the past two decades development of new control strategies has been largely based on modern and classical control theories. Modern control theory such as adaptive and optimal control techniques and classical control theory have been based mainly on linearization of systems [1]-[5].

In the application of such techniques, development of mathematical models is a prior necessity. However, such mathematical modeling which is largely based on the assumptions of linearization of systems might not reflect the true physical properties of the system. Even if complex mathematical models which do reflect precise input-output physical relations of the system can be built, the sensitivity of the parameters should be low in order to make the control system useful. This is rather difficult except for the case of linearized models.

Generally speaking, 'good' mathematical models for the approximation of the real physical properties of systems were statistical models [2], for example, auto-regressive (AR) model, auto-regressive moving-average (ARMA) model, lattice model, etc. These models have been developed by statisticians as well as engineers in the signal processing field and in many applications [3], such as, speech recognition, geological data analysis, image processing, etc., rather successful results have been obtained. In addition to simplicity for modeling, the statistical models can check the validity of the model from a viewpoint of frequency domain by using spectral analysis.

Thus, we can estimate what happens in a specific frequency band when we use statistical models such as in the case of frequency response analysis in classical control theory. These statistical approaches seem excellent but for control applications, such an approximation is not so precise. This is due to the fact that linear models are fitted to complex nonlinear systems. Furthermore, the methods are based on black box models and sometimes the variables or parameters have no physical meaning. This is one of the reasons why control engineers have not readily accepted such approaches [4].

Another popular control approach is adaptive control which has been discussed in the previous chapter [5]. In this case, we assume a mathematical model based on the physical phenomena and estimate the unknown parameters included in the mathematical model. Then a control law is determined to achieve some goals of the cost function by regarding the mathematical model as the true system. Such an adaptive control approach is also based on linear system theory. If some changes in the plant or environment occur, we must re-build the model and determine a new control law for the model. Thus, we must check manually whether the model is adequate to describe the real physical system or not.

Considering these factors, modern control theory was not very popular and rather difficult to be applied for real world problems except for some specific application area such as 'guidance control of satellites'. For example, in Japan about 84% of the control industry still uses the conventional PID controller [4].

From these facts, it seems that the specifications required for real applications of control theories are that the control algorithms should be simple enough to be implemented and to be understood. Perhaps these algorithms should also have the following properties such as learning ability, flexibility, robust, and nonlinearity. One of the reasons why fuzzy control has become so popular now is that it satisfies some of these factors.

Neural networks are currently being sought for control purposes. To this end, they have been proven to be successful in solving pattern recognition problems. Neural networks have the ability to learn from input-output functions, therefore, they provide simpler solutions to complex control problems. Furthermore, neurons are nonlinear elements and hence, neural networks are basically nonlinear systems which can be used to learn and solve nonlinear control problems which traditional and conventional control methods have no solution yet. Thus, in recent years, intelligent control in general has been quite readily acceptable for real control applications [5], [6].

4.2 Overview of Neuro-Control

Neuro-control is defined as the use of well-specified neural networks - artificial or natural - to emit actual control signals. Usually, when we talk about 'control', there is some form of intelligence associated with the term. We need to control themselves from doing unlawful things. In many types of religions, people are always reminded not to transgress themselves. Hence, they need to control themselves from transgression. When we try to control our emotions or some things, then there is an element of thinking which relates to some 'form' of intelligence. In engineering terms, when we design a controller to control a process or plant according to the desired set-points, it can be said that there is some process of thinking (though crudely) that is needed to be done by the controller to make the plant behave according to the desired values indicated. One characteristic of feedback control systems is that they can "think" in the sense that they can replace, to some extent, human operations. Norbert Wiener in his book "Cybernetics" [7] implied that all systems, living and mechanical, are both information and feedback control systems. He once implied that control, information, and neural science were regarded as a common subject under the banner of cybernetics. However, these disciplines have tended to go their own separate ways.

In the last decade, the backpropagation algorithm has been recognized as an efficient tool to realize the learning mechanism after its discovery. In almost every field of engineering, researchers began to actively engage in the applications of neural networks with the intention of finding better solutions to conventional methods. In the field of control, neural networks were found to be suitable for solving nonlinear and complex control problems where to this end conventional and traditional control methods have no practical solution yet. From the authors' point of view, there are several reasons that have motivated vast research interests in the application of neural networks for control purposes, as alternatives to traditional control methods, among which the main points are:

- Neural networks can be trained to learn any function [8], [9], [10] provided that enough information is given during the training process coupled with judiciously selected neural models. Thus, this self-learning ability of the neural networks eliminates the use of complex and difficult mathematical analyses which is dominant in many traditional adaptive and optimal control methods.
- The inclusion of semilinear sigmoidal activation function (or certain general nonlinear functions [11]) in the hidden neurons of multilayered neural networks offers nonlinear mapping ability for solving highly nonlinear control problems where to this end traditional control approaches have no practical solution yet. Perhaps, this is the most significant advantage of using neural networks from the control theoretical viewpoint.
- The requirement of vast *a priori* information regarding the plant to be controlled such as mathematical modeling is a prior necessity in traditional adaptive and optimal control techniques before they can be implemented [1], [12]. Due to the self-learning capability of neural networks such vast information is not required for neuro-controllers. Thus, neuro-controllers seem to be able to control under a wider range of uncertainty [13], [14].
- The massive parallelism of neural networks offers very fast multiprocessing technique when implemented using neural chips or parallel hardware's. American Neurologix Inc. has just developed a neural processor (NLX420) which are already available commercially [15].

- Damage to some parts of the neural network hardware may not affect the overall performance badly due to its massive parallel processing architecture [16], [17].

In recent years, there have been tremendous research articles and papers which have been reported in journals and conferences devoted to the topic of artificial neural networks in relation to the field of control science. This field of neuro-control which is becoming more important each day cannot be ignored to be a mere hyperbole anymore where it was once considered to be so by some distinguished researchers in the field of control. One of the first significant attempts to publish a book on neuro-control is the one edited by Miller *et al.* [18], entitled "Neural Networks for Control" which consists of a compilation of significant early contributions on neuro-control.

The article by Barto [19] in Chapter 1 of that book provides some relevant discussions of neuro-control architecture's for learning, identification, and control. In the same book Werbos [20] attempted to outline some classification of neuro-control applications though some may not agree. For example, his classification of 'supervised control' and 'direct inverse control' is rather confusing as 'supervised control' can be a subset of 'direct inverse control', and vice-versa. Furthermore, some may argue that his classification of the 'neural adaptive control' approach is not much different from the 'backpropagation-through-time' approach which is also confusing. Though many believe that it is still early to make such classifications, the attempt by Werbos [20] should be given some credit.

In the preceding part of this section, we attempt to classify some neuro-control approaches from our viewpoint which we will discuss in detail in this chapter. Perhaps to this end, one of the most significant surveys on neuro-control in perspective is the paper by Hunt *et al.* [21]. In their paper, a view regarding the parallelism between the field of control science and neural networks is very relevant indeed. In their survey, they also covered almost all of the neural paradigms that have been applied by researchers for identification and control purposes. In this overview we will discuss briefly some of the more important architectures of neuro-control that have been proposed (mainly the ones that are much cited in the literature).

4.2.1 Neuro-control Approaches

Perhaps one of the first significant works on neuro-control is the one of Widrow and Smith [22]. They showed that their ADALINE is capable of balancing a cart-pole system by copying an existing human controller which is a form of 'supervised control'. Albus [23], [24] proposed a new approach to manipulator (robotic-arm) control using a form of a look-up table neural network known as CMAC (cerebellar model articulation controller). Miller III *et al.* [25], has applied the CMAC in controlling robots and other applications in real-time at the University of New Hampshire. In 1983, Barto *et al.* [26] proposed an adaptive learning system consisting of (which they referred to as) a single associative search element (ASE) and a single adaptive critic element (ACE) applied to balance a cart-pole system as shown in the simplified block diagram as in Fig. 4.2.1. Werbos [19] classified this approach as 'adaptive critic' and several variants of this credit-assignment approach have been independently studied and applied by others (Gaines and Andreae [27], Witten [28], Werbos [29], Holland [30], Anderson [31], [32]).

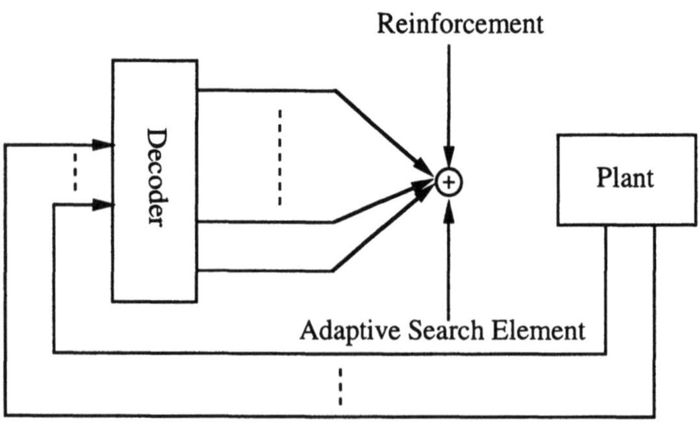

Fig. 4.2.1. Adaptive critic method of Barto, Sutton, and Anderson [26].

The success of the backpropagation algorithm to train multilayered networks started an explosion on the application of neural networks for control purposes. Various neuro-control schemes based on the backpropagation algorithm were proposed. Perhaps, one of the most widely applied neuro-control schemes is the direct inverse model neuro-control approach. The popularity of this approach is its simplicity. Once a neural network has been trained to learn the inverse of a plant it can then be configured to control the plant directly .

The idea of this scheme was perhaps borrowed from the traditional self-tuning control scheme [1] where the correct control signal to the plant can be obtained from its inverse mathematical model by giving the desired plant output. Widrow and Stearns's book [33] "Adaptive Signal Processing" showed some examples on inverse model control of linear systems using neural networks. Other applications of this approach include the works of Kuperstein [34], Kuperstein and Rubinstein [35], Srinivasan *et al.* [36], and Sanner and Akin [37].

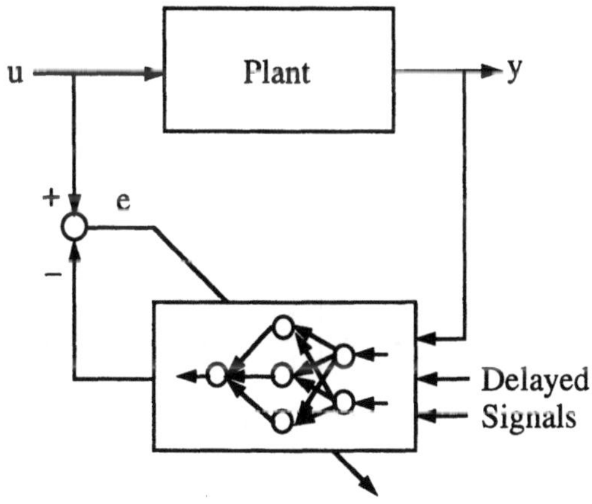

Fig. 4.2.2. Generalized learning or direct inverse control architecture. The neural network model is trained off-line to learn the inverse model of the plant. After learning it is then configured to control the plant directly.

The paper by Psaltis *et al.* [38] is perhaps one of the most cited papers in the field of neuro-control. They proposed two simple but effective methods of training the neural networks; two inverse plant models using the backpropagation algorithm referred to as generalized and specialized learning architectures. In the generalized learning architecture as shown in Fig. 4.2.2, the network is trained off-line using patterns obtained from the plant's open-or closed-loop characteristics. This method is similar to batch learning as in pattern recognition problem. Once trained the network is configured as a direct controller to the process similar to a conventional feedback control system. Using the specialized learning architecture (Fig. 4.2.3), the network is trained in an on-line way (or 'goal-directed') where the performance error is backpropagated through the network at every sample. However, due to the location of the plant the Jacobian of the plant is required. In order to avoid this, the elements of the Jacobian may be approximated by their signs which are the orientations of the control parameters influencing the outputs of the plant as studied by Saerens and Soquet [39].

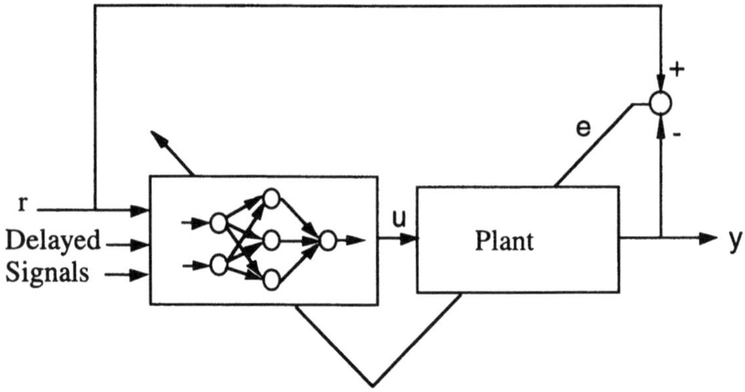

Fig. 4.2.3.Specialized learning architecture. The neural network model is trained on-line to learn the inverse model of the plant by backpropagation of the performance error.

Another inverse model of neuro-control on-line learning scheme (though not direct) utilizing the backpropagation algorithm is the popular 'feedback-error learning' scheme of Kawato *et al.* [40], [41]. In this scheme as shown in Fig. 4.2.4, the neural network is configured in parallel to a conventional feedback controller (which we will refer to as a parallel neuro-control scheme) and the network is trained on-line by repeating the desired trajectory cycles where the feedback error is backpropagated through the network. Convergence is achieved when the neural network has learned the inverse of the plant and it then takes control of the plant and eliminates the significance of the feedback controller.

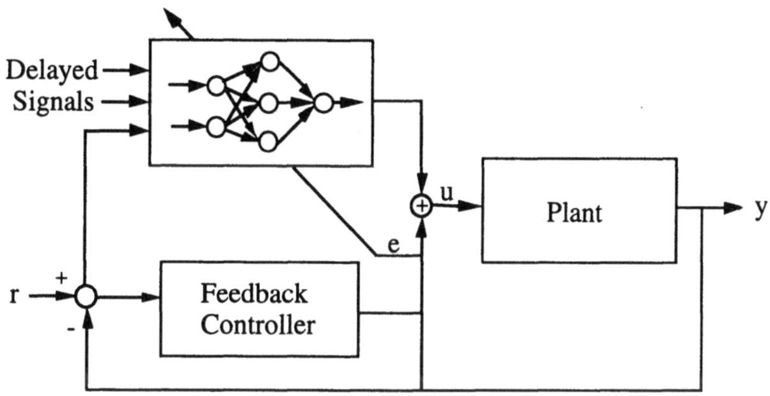

Fig. 4.2.4. Fccdback-crror lcarning architecture. The neural network model is trained on-line to learn the inverse model of the plant by backpropagating the output of the feedback controller.

This method of control can be thought of similar to the way when a person first try to play tennis (as an example). A novice tennis player would not be able to hit the ball in a correct manner at first and would do every effort he has to hit the ball as accurately as possible (before learning perhaps similar to conventional control) but after some practice he would be able to control his swing to hit the ball in a more correct manner (superior control after learning). Due to the training method of repeating the trajectory cycle over several hundred thousand of cycles, this feedback-error learning method is practically limited to certain types of systems only, i.e., fast-acting systems such as robotic-arms and servo-motors.

Another notable neuro-control architecture utilizing the backpropagation algorithm is the scheme which has been proposed independently by Jordan [42], and a more recent paper by Jordan and Rumelhart [43], Narendra and Parthasarathy [44], Nguyen and Widrow [45], and Werbos [20]. Werbos [20] classified this method as 'backpropagation-through-time'. Though Jordan and Rumelhart [43] claimed that they were the first to develop this approach, the architecture is nothing new in the eyes of control scientists as it resembles many traditional adaptive control structure which is called indirect adaptive control [1]. In this scheme two neural networks are used to control the plant as shown in Fig. 4.2.5. The first neural network is used as an emulator and the other as a controller. The emulator neural network may be trained off-line using the generalized learning architecture of Psaltis *et al.* [38] or even on-line by injecting random inputs to learn the forward plant dynamics.

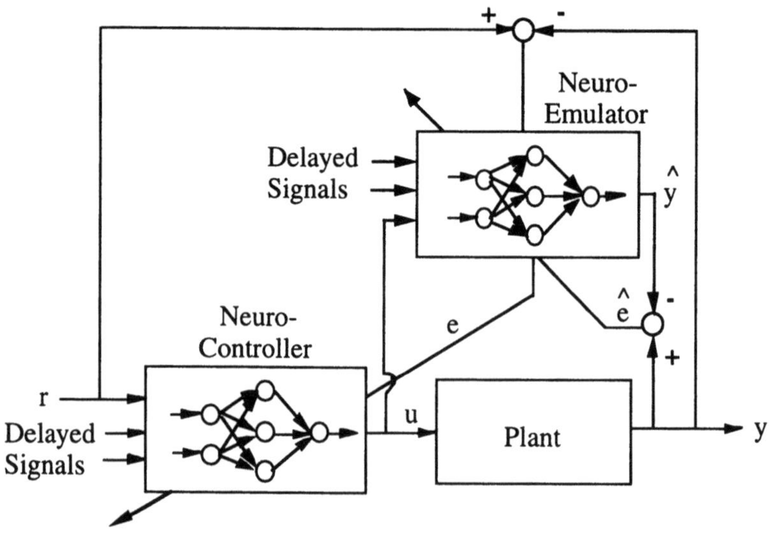

Fig.4.2.5. Emulator and controller neuro-control scheme or "backpropagation-through-time". The neuro-controller is trained by the inverse plant model and the neuro-emulator is trained by the forward plant model. The neuro-controller can be trained on-line by backpropagation of the performance error through the neuro-emulator.

This architecture allows a more accurate approach of training the neuro-controller on-line as the performance error can be backpropagated through the emulator at every sample. In Chapter 5, we also apply this architecture for on-line learning of the neuro-controller for two applications.

Our discussions on neuro-control so far have mainly been on control of plants, though the last scheme mentioned did have some form of plant identification using the neuro-emulator. However, the neuro-emulator in this case is merely used as means to backpropagate the performance error to get the equivalent error at the output of the neuro-controller. Due to their learning ability, neural networks can also be used for system identification which has been one of the major factors for the development of many traditional adaptive controllers. There has also been substantial research in applying neural networks for plant identification in which we do not wish to elaborate in this thesis. The following papers by Chen *et al.* [46], [47], Narendra and Parthasarathy [44], Bhat *et al.* [48], Chu *et al.* [49], Billings *et al.* [50], and Sjoberg and Ljung [51] are some of those that can be referred to as the application of neural networks for system identification.

From our viewpoint most of the neuro-control schemes developed until now are based on the following five design approaches:

(1) Series control scheme; the neural network directly learns the mapping from desired reference signals to the control inputs which yields these signals.

(2) Parallel control scheme; a neural network is used to compensate the control signal which is provided by a conventional controller such that the plant output can track the desired output as close as possible.

(3) Self-tuning control scheme; a neural network tunes the control parameters included in a conventional controller such that the plant output follows the desired output signal as much as possible.

(4) Emulator and controller scheme or backpropagation-through-time; it maximizes some measure of utility or performance over time, but cannot efficiently account for noise and cannot provide real-time learning for slow convergence.

(5) Adaptive critic scheme; it approximates dynamic programming, that is, approximates optimal control over time in noisy and nonlinear environments.

Except for the last scheme, the above neuro-control design methods will be explained in more detail in the following sections which are all based on the backpropagation algorithm.

4.2.2. General Control Configuration

The configuration of this control scheme is shown as in the block diagram of Fig. 4.2.6. We can adjust two parts to achieve the desired control result such that the plant output follows a reference signal as much as possible. The two parts are feedforward controller (abbreviated as FFC) and feedback controller (abbreviated as FBC) which are shown by the acclivity arrows which mean the variable parts.

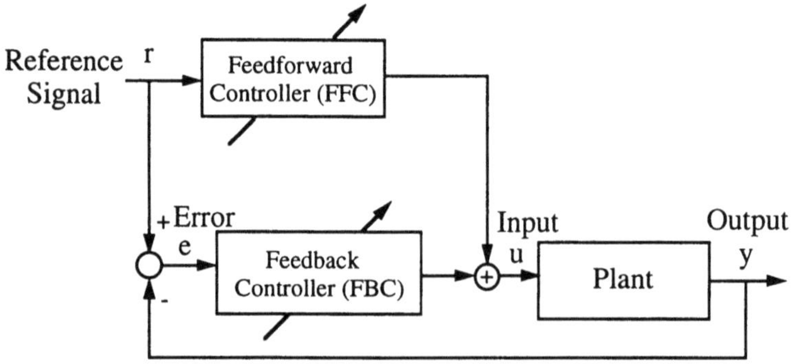

Fig. 4.2.6. Block diagram of a general control scheme in the case of no disturbance.

The FBC is used to stabilize the closed-loop system based on the error signal which is the difference between the reference signal and the plant output. If the plant has a large time-delay, an effect of control input will appear in the plant output at some later time and the plant output is fedback and compared with the reference signal. The error is used to determine the next control input. Thus, such a delay may cause the closed-loop system to be rather unstable.

In the case of no disturbance, the FFC does not fully guarantee that the plant output follows the reference signal but the decision is delivered quickly to the plant. Hence, this makes the system response very fast. Roughly speaking, the FBC will stabilize the system while the FFC will make the system to respond veru quickly. Therefore, if we wish for the plant output to follow quickly the reference signal, the FBC can be adopted. In the following sections, we will explain how to build the FFC or FBC blocks by using neural networks.

4.3 Series Neuro-Control Scheme

The series neuro-control scheme is shown in Fig. 4.3.1. This type of neuro-control is also referred to as direct inverse control, which means that the neural network may realize the inverse dynamics of the plant.

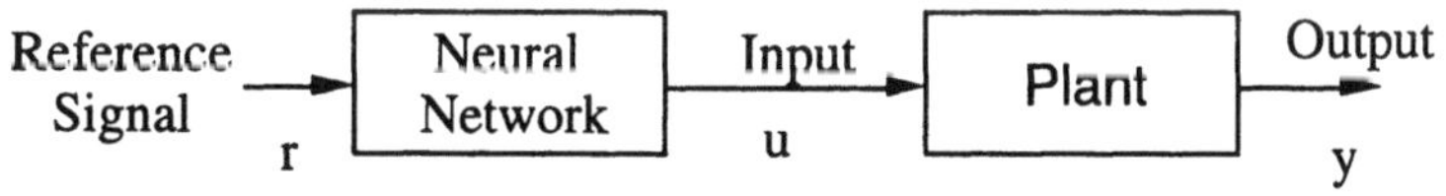

Fig. 4.3.1. Block diagram of a series neuro-control scheme.

Thus, if we denote the input/output mapping of the plant by $f_p(u)$, that is,

$$y = f_p(u), \tag{4.3.1}$$

the neural network realizes the inverse mapping as

$$u = f_p^{-1}(y)$$

where f_p^{-1} denotes the inverse mapping of f_p.

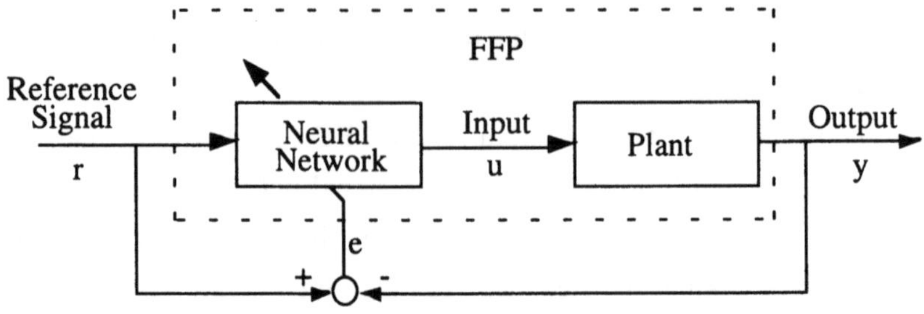

Fig. 4.3.2. Specialized inverse mapping neuro-control where the part enclosed by dotted lines is a feedforward path (FFP).

Thus, if we give the reference signal r to the neural network, then the plant output y becomes r, that is,

$$y = f_p(u) = f_p(f_p^{-1}(r)) = r \tag{4.3.2}$$

where u is the plant input corresponding to the reference signal r and y is the plant output to u.

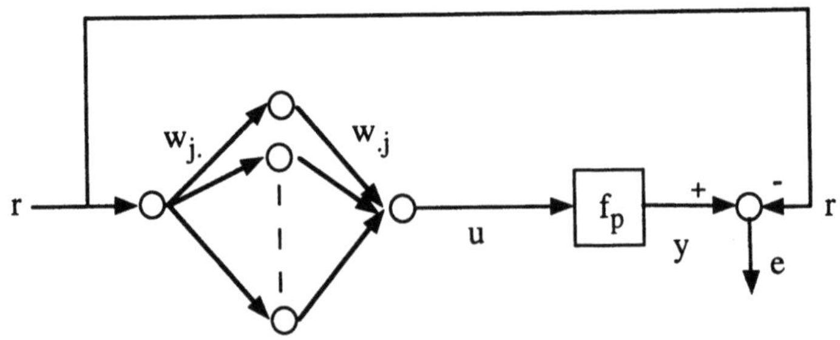

Fig. 4.3.3. Internal structure of the neuro-controller of Fig. 4.3.2 showing where w.j and wj. correspond to w_{kj} and w_{ji} in Section 2.2.1.

In order to realize the configuration of Fig. 4.3.1, we can use the following two types. One is called the specialized inverse mapping given by Fig. 4.3.2. In Figure 4.3.2 the neural network is trained such that the squared error is minimized. If the error becomes almost equal to zero, then the neural network realizes the inverse dynamics of the plant. The part enclosed by the dotted lines, which is a feedforward path (abbreviated as FFP) can be regarded as a kind of neural networks whose output layer is hard-wired. In this case, we will show that the system's Jacobian is required in order to use the backpropagation algorithm.

The simplest internal structure of the neural network of Fig. 4.3.2 can be given by Fig. 4.3.3 where one neuron in the input and output layers is used. In Fig. 4.3.3 input to the neural network is only r and the output of the neural network is only u. For simplicity, we denote the input/output relation of the plant by f_p in which the output is desired to be r as close as possible after learning. Thus, the error function to be minimized becomes

$$E = \frac{1}{2} e^2 \tag{4.3.3}$$

$$e = r - y \tag{4.3.4}$$

$$y = f_p(u). \tag{4.3.5}$$

The plant input u is described as

$$u = f(net_{.}) \tag{4.3.6}$$

$$net_{.} = \sum_j w_{.j} O_j + \theta_{.} \tag{4.3.7}$$

$$O_j = f(net_j) \tag{4.3.8}$$

$$net_j = w_j . r + \theta_j \tag{4.3.9}$$

where $w_{\cdot j}$ and $\theta_{\cdot}$ are the connection weights and a threshold, respectively, at the output layer of the neural network and $w_{j\cdot}$ is a connection weight at the hidden layer. Using the steepest descent principle, we have

$$\Delta w_{\cdot j}(t+1) = -\eta \frac{\partial E}{\partial w_{\cdot j}} + \alpha \Delta w_{\cdot j}(t). \tag{4.3.10}$$

Using the chain rule of derivative, we have

$$\frac{\partial E}{\partial w_{\cdot j}} = \frac{\partial E}{\partial net_{\cdot}} \frac{\partial net_{\cdot}}{\partial w_{\cdot j}}. \tag{4.3.11}$$

From (4.3.7) we have

$$\frac{\partial net_{\cdot}}{\partial w_{\cdot j}} = O_{j\cdot}. \tag{4.3.12}$$

Defining $\delta_{\cdot}$ as

$$\delta_{\cdot} = -\frac{\partial E}{\partial net_{\cdot}} \tag{4.3.13}$$

and using the chain rule again for $\delta_{\cdot}$,

$$\begin{aligned} \delta_{\cdot} &= -\frac{\partial E}{\partial e}\frac{\partial e}{\partial y}\frac{\partial y}{\partial u}\frac{\partial u}{\partial net_{\cdot}} \\ &= e\frac{\partial y}{\partial u}\frac{\partial u}{\partial net_{\cdot}} \\ &= e f'_p(u)\, f'(net_{\cdot}) \\ &= e f'_p(u)\, u(1-u). \end{aligned} \tag{4.3.14}$$

Thus, we have the following relation.

$$\Delta w_{\cdot j}(t+1) = \eta\, \delta_{\cdot} O_{j} + \alpha \Delta w_{\cdot j}(t) \tag{4.3.15}$$

$$\delta_{\cdot} = e f'_{p}(u) u(1-u) . \tag{4.3.16}$$

Similarly, for $\Delta w_{j\cdot}$ we have

$$\Delta w_{j\cdot}(t+1) = \eta \delta_{j} r + \alpha \Delta w_{j\cdot}(t) \tag{4.3.17}$$

and

$$\delta_{j} = -\frac{\partial E}{\partial net_{j}} = -\frac{\partial E}{\partial net_{\cdot}} \frac{\partial net_{\cdot}}{\partial O_{j}} \frac{\partial O_{j}}{\partial net_{j}} \tag{4.3.18}$$

$$= \delta_{\cdot} w_{\cdot j} f'(net_{j}) = \delta_{\cdot} w_{\cdot j} O_{j}(1 - O_{j}) . \tag{4.3.19}$$

Therefore, we have the following algorithm:

Step 1. Set the initial values of $w_{\cdot j}$, $w_{j\cdot}$, θ_{j}, $\theta_{\cdot}$, η, and α.
Begin $t = 0$ and go to Step 2.

Step 2. Apply the desired output r(t) as the input to the neural network and calculate u(t) and y(t). Then compute $\delta_{\cdot}$ by the formula:

$$\delta_{\cdot} = u(t)(1-u(t))e(t)f'_{p}(u(t))$$

$$e(t) = r(t) - y(t).$$

Step 3. Change the connection weights by

$$\Delta w_{\cdot j}(t+1) = \eta \delta_{\cdot} O_{j} + \alpha \Delta w_{\cdot j}(t) .$$

Step 4. Calculate δ_j by

$$\delta_{j} = w_{\cdot j}\, \delta_{\cdot} O_{j}(1-O_{j}) .$$

Step 5. Change the connection weights by

$$\Delta w_{j\cdot}(t+1) = \eta \delta_{j} r(t) + \alpha \Delta w_{j\cdot}(t) .$$

Step 6. t → t+1 and go to Step 2.

In real applications there are some problems. One is the input/output mapping by the neural network of Fig. 4.3.3 which is not a dynamic but rather a static mapping from u to y. In the control application field dynamic mappings are more usual and hence, the neural network must provide many past inputs to the plant on-line or the past inputs must be stored. In the former case, the outputs of the neural network are not only u(t) but also u(t-m), m=1, 2, ..., q and in the latter case, the inputs of the neural network become r(t - i), i = 1, 2, ..., k.

Furthermore, we can provide more information to the input of the neural network in order to realize a fast convergence of learning. If we know some factors which can improve the control signal, it is better to use such information as the input vectors of the neural network.

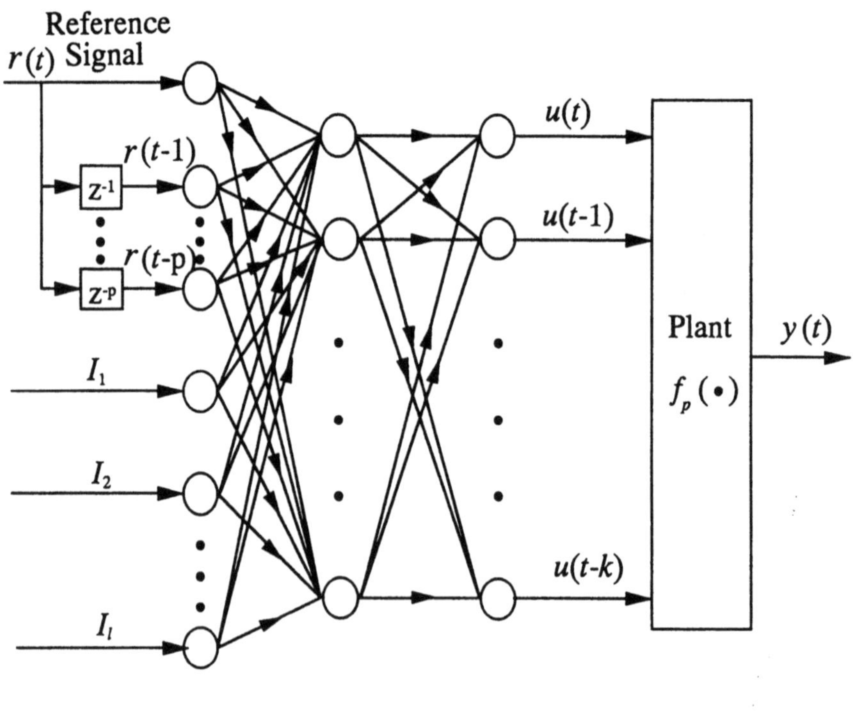

Fig. 4.3.4. General configuration for a series type neuro-controller.

Therefore, a general configuration for a series type neuro-controller can be shown as in Fig. 4.3.4 where $I_1, I_2, ..., I_l$ denote additional informations which may affect the control signal. However, if we do not know the time-delay q and p precisely, we cannot use such a structure. In this case we may use the estimated values of q and p.

Using the control configuration of Fig. 4.3.4, the same methodology as described above can be used to train the neural network. Thus, if we can correspondingly equate the following;

$$\begin{aligned}
&w_{\cdot j} \to w_{kj} \\
&w_{j\cdot} \to w_{ji} \\
&\delta_{\cdot} \to \delta_k \\
&f_p(u(t)) \to f_p(u(t),(t-1),\dots,u(t-q))
\end{aligned} \tag{4.3.20}$$

then we can apply Steps 1 - 6 as described to the neuro-controller.

We must note that the algorithm stated above is based on the backpropagation algorithm with sigmoid functions as the activation functions. This means that all control inputs are restricted within 0 and 1. Thus, if we use only positive values as control inputs, then we must multiply the control signal u by some positive constant C or if we need negative values only as control inputs, then negative constant C should be multiplied to the control signals. To determine C we can include additional layer. More simply, if we take the identity function as the activation function at the output layer, then we can choose any value which is positive or negative.

In this case, Step 2 should be replaced as follows:

Step 2: Apply the desired output r(t) as the input to the neural network and calculate u(t) and y(t). Then compute $\delta_{\cdot}$ by the formula

$$\delta_{\cdot} = e(t) f'_p(u(t))$$

$$e(t) = r(t) - y(t)$$

In the above algorithm, the only difficulty is to calculate the Jacobian of the system $f_p'(u(t))$ at Step 2. Generally, f_p' is not known and hence, it should be approximated. One way is to use a numerical approximation such that

$$f'_p(u(t)) \approx \frac{\Delta f_p(u(t))}{\Delta u(t)}$$

where

$$\Delta f_p(u(t)) = f_p(u(t-1)) - f_p(u(t-2))$$
$$\Delta u(t) = u(t-1) - u(t-2) .$$

Another approach is to simulate the plant by using another neural network which maps the input/output relation of the plant. Thus, such a neural network is used as a system identifier. In neural network field, this is called an emulator instead of a system identifier. Then using the emulator, the systems Jacobian can be calculated as if the emulator were the true plant. That is to say,

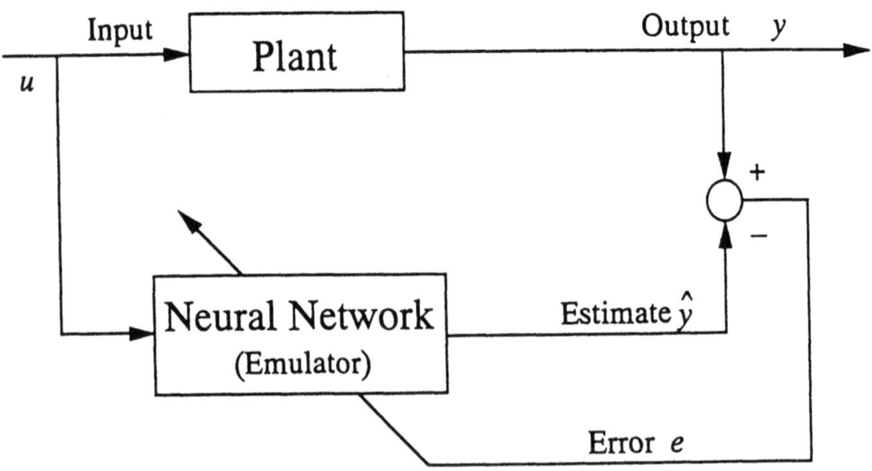

Fig. 4.3.5. A simple training configuration for developing a neuro- emulator.

$$f_p'(u(t)) \approx \frac{\partial \hat{f}_p(u(t))}{\partial(u(t))}$$

where $\hat{f}_p$ denotes an input/output mapping by the emulator. A simple training procedure for the emulator is described as in Fig. 4.3.5.

In Fig. 4.3.5, the neural network is trained such that prediction error $e = y - \hat{y}$ is minimized where $\hat{y}$ is an output of the emulator which denotes the predicted estimate of y. In a more detail manner, the structure given by Fig. 4.3.5 can be modified as Fig. 4.3.6. If we know the orders of time-delays of the input and output signals, then those values are used to determine p and q in Fig. 4.3.6.

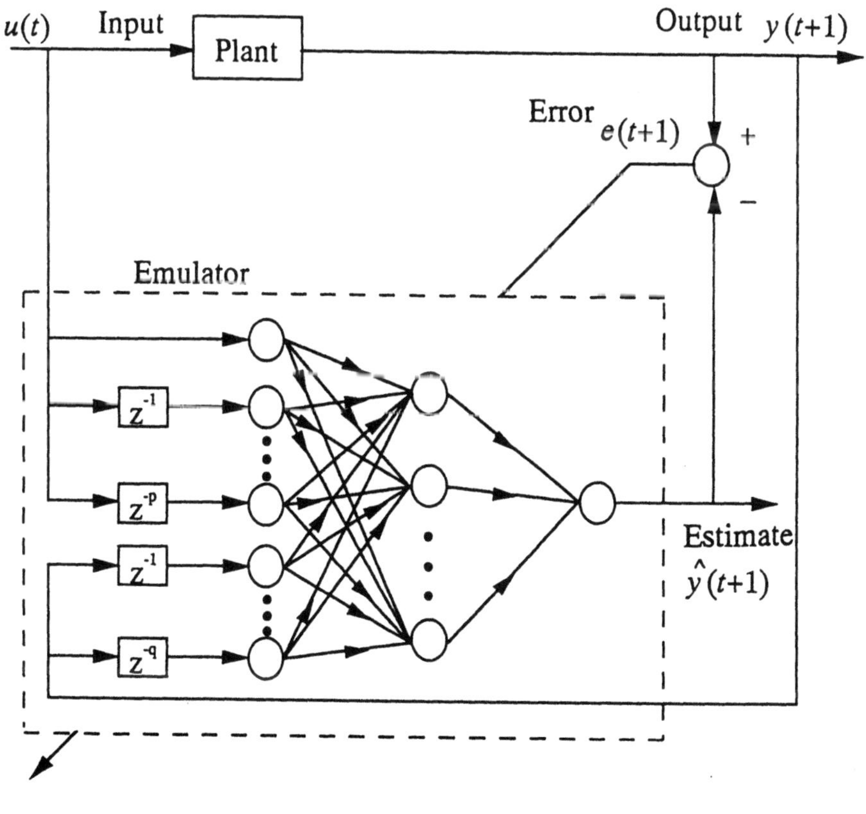

Fig. 4.3.6. Internal structure of a neuro-emulator where p and q are the orders of time-delays of input and output signals, respectively.

In order to speed up the convergence, we can use another model as an emulator, which is shown in Fig. 4.3.7. In this figure, a simple mathematical model or a physically precise model is used as a conventional model even if it is complex and nonlinear. If there is discrepancy between an actual output and a model output, the neural network will adjust itself to minimize the error. The model output is denoted by $\hat{y}_1$ and the output of the neural network is $\hat{y}_2$. Their sum is the predicted value $\hat{y}$ of the actual output y and error between y and $\hat{y}$ is denoted by ε.

If ε is not equal to zero, the neural network will learn to decrease the error ε. Thus, the emulator is regarded as the total sum of conventional model and neural network, but only the neural network can be adjusted. The modeling method is acceptable for control engineers since the conventional model may be insufficient to provide the perfect prediction of the actual output.

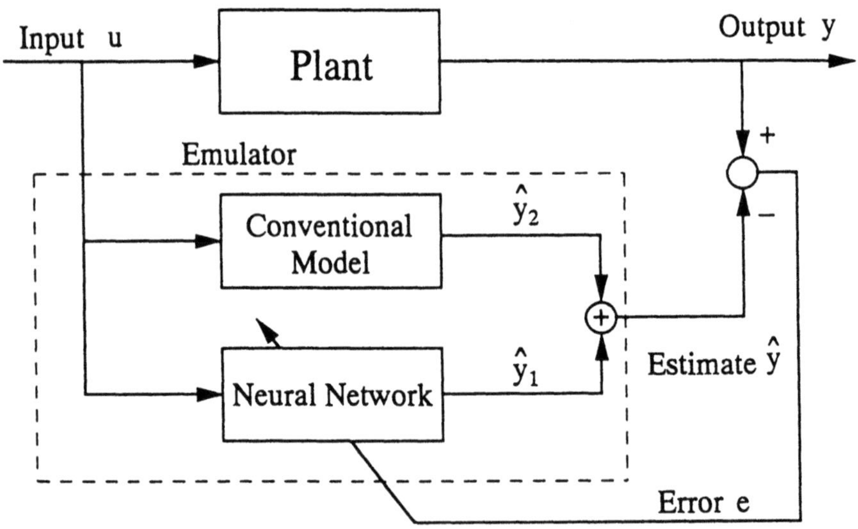

Fig. 4.3.7. An alternative modeling method by using neural networks.

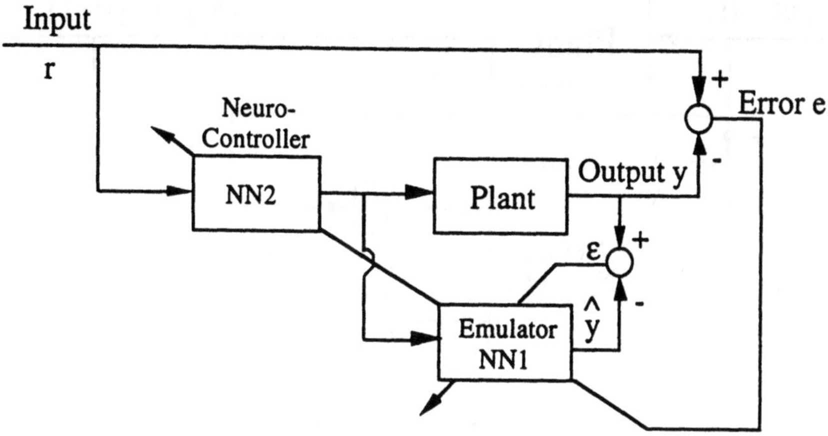

Fig. 4.3.8. Overall configuration of inverse dynamical neuro-controller where emulator NN1 is given by Figs. 4.3.6 and 4.3.7 and neuro-controller NN2 is constructed by Fig. 4.3.4.

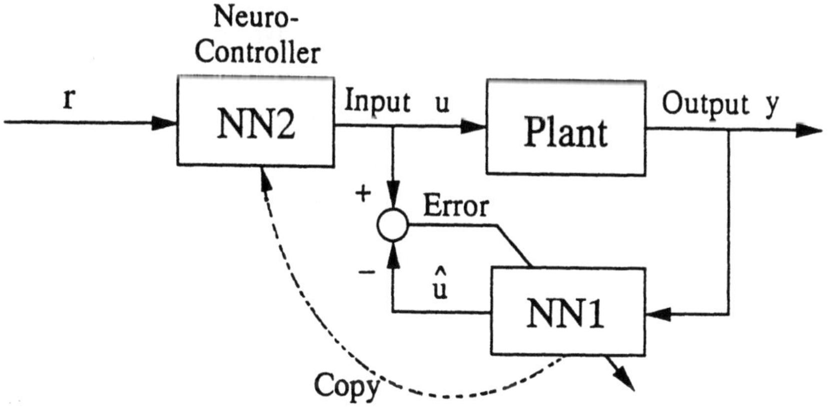

Fig. 4.3.9. Another architecture of the inverse mapping.

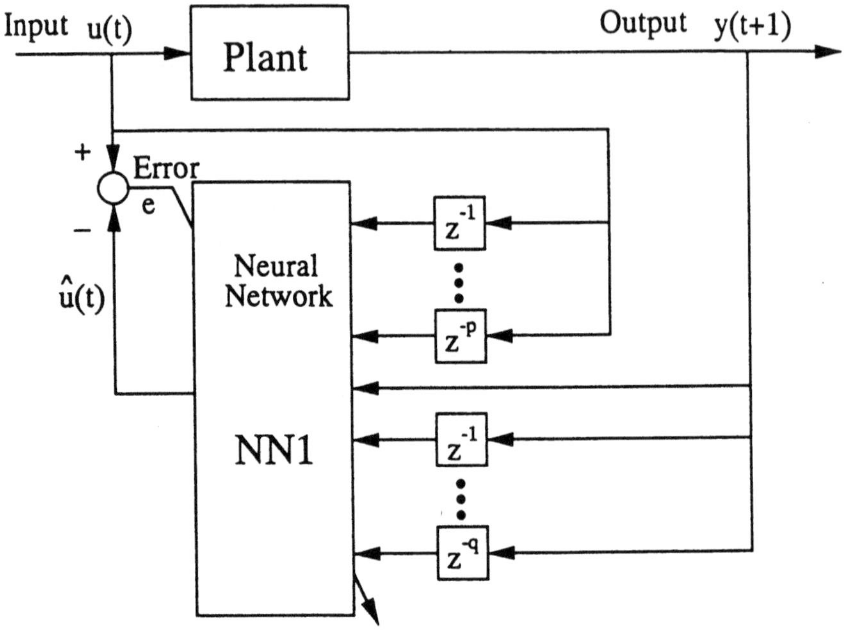

Fig. 4.3.10. Direct inverse modeling method of Fig. 4.3.9.

In our experience, conventional models such as static model, auto-regressive (AR) model, auto-regressive moving-average (ARMA) model, linear vector equation etc. are recommended since they can extract a long term trend by linear regression and the neural network can realize a nonlinear relation. Therefore, the total configuration of the neural network control scheme is given as in Fig. 4.3.8 where NN1 denotes a neural network emulator or simply neuro-emulator and NN2 denotes the neuro-controller which realizes the inverse dynamical mapping.

Another way to realize the inverse dynamics by a neural network is given by Fig. 4.3.9 where NN1 is used for training the inverse mapping and NN2 is copied from NN1 after NN1 has been trained. This is the architecture proposed by Psaltis *et al.* [38] as described in Section 4.2.1. The neural network NN1 is realized as shown in Fig. 4.3.10 where p and q correspond to time-delays of input and output signals, respectively.

4.4. Extension of Series Neuro-Control Scheme

Although neuro-controllers based on multilayered neural networks are popular in the non-conventional branch of learning or adaptive control, practical applications are strongly limited by the long training time that they require. This section addresses the question of how to perform on-line training of multilayered neuro-controllers in an efficient way in order to reduce the training time.

At first, based on multilayered neural networks, control configuration using neuro-emulators and neuro-controllers are described. The neuro-controller will learn the inverse dynamics of the plant from randomly chosen initial weights. Here, in order to describe several efficient on-line training methods, based on performing multiple updating operations during each sampling period, we will present the basic control configurations again. One method, the direct inverse control approach, is effective for small adjustments of the neuro-controller when it has been already trained. Another, the predicted output error approach, directly minimizes the control error and greatly improves convergence of the controller. Simulation results using a simple plant model of a water bath show the effectiveness of the proposed neuro-control structures and training methods.

4.4.1 Some Discussions on On-Line Learning

Artificial neural networks are mathematical systems designed to deliberately employ principles on which biological nervous systems are believed to be based. By embodying such principles, neural network modelers expect to be able to emulate the information processing capabilities of biological neural systems to some extent [5]. The recent efforts in applying neural networks to the control of dynamical processes resulted in the fledgling, but very promising, field of neuro-control which can be thought of as a non-conventional (connectionist) branch of adaptive control theory. The neuro-control appeal for control engineers can be primarily explained by three reasons:

(1) Biological nervous systems are living examples of intelligent adaptive controllers.

(2) Neural networks are essentially adaptive systems able to learn how to perform complex tasks.

(3) Neuro-control techniques are believed to be able to overcome many of the difficulties that conventional adaptive control techniques suffer when dealing with nonlinear plants or plants with unknown structure.

Generally, the training of neural networks for control can be performed on- or off-line, depending on whether they execute useful work or not while learning is taking place. Although off-line training is usually straightforward, conditions for assuring good generalization of the neural networks through the control space are difficult to attain, which makes on-line training always necessary in control applications. In fact, the training should ideally occur exclusively on-line, with the neural networks learning at high speed from any initial set of weights.

Neuro-controllers based on neural network structures, have been proposed to learn the inverse dynamics of the control plant from the observation of the plant input-output relationship through time. Although such neuro-controllers have proven to be effective in controlling complex systems, the usually very long time required for training makes controllers obtained via conventional techniques preferable in many practical applications. The slow convergence of neuro-controllers implies poor control performance and robustness, especially, during the initial stages of training. In order to make neuro-control a viable alternative to industrial control of processes, there is a pressing necessity for efficient on-line training algorithms for neuro-controllers.

This section first presents some general neuro-control structure based on multilayered neural networks. Only a little qualitative knowledge about the plant is necessary. Basic neuro-control configurations are briefly discussed, and though the neuro-controllers perform quite well using these training configurations, rather long training time is necessary. In conventional adaptive systems in which one updating operation takes place for each sampling period, increasing the sampling rate seems to be a natural step toward improving performance. However, in many practical cases the sampling rate cannot exceed some limit, either by physical or practical (technical) constraints. Considering such a limitation, new on-line training methods, based on efficient use of plant input-output data and distinction between sampling and learning frequencies, are proposed in order to reduce neuro-controller training time and described in algorithmic form [52]. Simulation results attest the effectiveness of the proposed neuromorphic control structures and training methods.

4.4.2 Neuromorphic Control Structures

4.4.2.1 Problem Statement

Consider the discrete-time single-input single-output (SISO) process

$$y(t+1) = f_p[y(t), y(t-1),..., y(t-\bar{q}+1), \\ u(t), u(t-1),..., u(t-\bar{p})] \tag{4.4.1}$$

where y denotes the output, u is the input, t is the discrete-time index, $\bar{q}$ and $\bar{p}$ are nonnegative integers, and $f_p(.)$ is some function. In many practical case, the plant input is limited in amplitude, i.e., there exist u_m and u_M such that, for any t,

$$u_m \leq u(t) \leq u_M . \tag{4.4.2}$$

Here, the task is to learn how to control the plant described in (4.4.1) in order to follow a specified reference r(t), minimizing some norm of the error e(t) = r(t) - y(t) through time. It is assumed that the only available qualitative *a priori* knowledge about the plant is that q and p, which are, respectively, estimates for $\bar{q}$ and $\bar{p}$ in (4.4.1), are known. In other words, it is assumed that a rough estimate of the order of the plant to be controlled is given.

4.4.2.2 Multilayered Neural Networks for Control

Although several neural network architectures have been applied to robotic and process control, most of the incipient neuro-control literature concentrates on multilayered neural networks. These neural networks are particularly attractive to control engineers due to the following basic reasons:

(1) They are essentially feedforward structures in which the information flows forward, from the inputs to the outputs, through hidden layers. This characteristic is very convenient for control engineers, used to work with systems represented by blocks with inputs and outputs and is not present in recurrent networks, e.g., Hopfield's model [53], in which bi-directional nodes appear and there are no true inputs and outputs.

(2) Those with as few as one hidden layer using arbitrary sigmoidal activation functions are able to perform any nonlinear mapping between two finite-dimensional spaces to any desired degree of accuracy, provided that there are enough number of hidden units (neurons). In other words, multilayered neural networks are very versatile mappings of arbitrary precision. In control, usually many of the blocks involved in a control system can be viewed as mappings and, therefore, can be emulated by multilayered neural networks with inputs and outputs.

(3) The basic algorithm for learning in multilayered neural networks, the backpropagation algorithm, belongs to the broad class of steepest descent (gradient) methods largely applied in optimal control, and is, therefore, familiar to control engineers.

Points (1), (2), and (3) indicate that multilayered neural networks can be thought of as blocks with a learning or mapping ability. Based on such mapping ability, two general neuro-control structures can be proposed; a plant emulator and a controller.

4.4.2.3 Neural Plant Emulator (PE)

Given the estimates p and q, a multilayered neural networks with $m = p + q + 1$ inputs and single output can be used for emulating $f_p(.)$ in (4.4.1). Denoting the mapping performed by the plant emulator by $\varphi_E(.)$, and its output by y_1, we have

$$y_1 = \varphi_E(x_E) \tag{4.4.3}$$

where x_E is an m-dimensional vector.

For $x_E(t) = [y(t),\ldots,y(t-q+1),\ldots, u(t),\ldots,u(t-p)]^T$, the emulator is trained in order to minimize a norm of the emulation error y(t + 1) - y_1. The PE is illustrated in Fig. 4.3.6.

4.4.2.4 Neuro-Controller

Assume that the plant in (4.4.1) is invertible, i.e., there exists a function g(.) such that

$$u(t) = g[y(t+1), y(t), \ldots, y(t-q+1), u(t-1), u(t-2), \ldots, u(t-p)]. \tag{4.4.4}$$

Consider again a multilayered neural networks with m-dimensional input vector x_c, single output u_1, and an input-output relationship briefly represented by

$$u_1 = \varphi_C(x_c) \tag{4.4.5}$$

where φ_C (.) denotes the input-output mapping of the multilayered neural networks. If the output of φ_C (.) approximates the output of g(.) for corresponding inputs, the multilayered neural networks can be thought of as a controller in the feedforward control path. At instant t, the input to the plant can be obtained from (4.4.5) by setting

$$x_c(t) = [r(t+1), y(t), \ldots, y(t-q+1), u(t-2), \ldots, u(t-p)]^T \tag{4.4.6}$$

where the reference r(t + 1) was used instead of the unknown y(t +1). After enough training of the neuro-controller, if the output error e(t) is kept small, it is possible to use

$$x_c(t) = [r(t+1), r(t), \ldots, r(t-q+1), u(t-1), \ldots, u(t-p)]^T \tag{4.4.7}$$

emphasizing the feedforward nature of the neuro-controller. This basic neuro-controller configuration is illustrated in Fig. 4.3.1 which we have referred as the series neuro-control scheme.

4.4.3 Training Configurations

The controller's training signal in Fig. 4.3.8 provides the information necessary for the neuro-controller to learn the inverse dynamics of the plant in such a way that an error function J defined on the plant output error e(t) = r(t) - y(t) is minimized. In order to enable learning based on the steepest descent (gradient) method, it is necessary to compute the derivative of the error function J with respect to the output of the neuro-controller, i.e., $\delta = -\frac{\partial J}{\partial u_1}$. The knowledge of δ suffices for updating the weights of the neuro-controller via backpropagation. In order to explicitly explain this efficient on-line training scheme for the multilayered neural networks, we will again briefly describe the basic neuro-control configurations in the following sections.

4.4.3.1 Direct Inverse Control

The direct inverse control configuration or general learning, depicted in Fig. 4.3.10, is an off-line training scheme. However, since this control scheme relies on the neuro-controller generalization ability, it should not be used alone as the only training scheme. On the other hand, training data for supervised learning can be obtained in a straightforward way. At time t + 1, for $x'_c(t) = [y(t + 1), ..., y(t + 1 - q), u(t - 1), ..., u(t - q)]^T$, the neuro-controller can be updated in order to minimize an error function J^u defined as a function of the difference $u(t) - u_1(t)$, where $u_1(t) = \varphi_c [x'_c (t)]$ and, therefore, the term $\delta^u = \partial J^u / \partial u_1$ can be easily calculated, enabling backpropagation. For example, defining

$$J^u(t) = \frac{1}{2}\left(u(t) - u_1(t)\right)^2 \qquad (4.4.8)$$

the expression for δ^u becomes merely

$$\delta^{u}(t) = u(t) - u_1(t) \tag{4.4.9}$$

where $\delta^{u}(t)$ denotes δ^{u} at time t.

4.4.3.2. On-Line Specialized Learning

This configuration is shown in Fig. 4.2.3. Learning is performed on-line and the error function J is based on the plant output error, as in many typical control optimization problems. The problems are that the exact calculation of δ (sensitivity of J with respect to the output of the neuro-controller) requires knowledge of the Jacobian of the plant. For example, a straightforward error function and corresponding δ term can be given by

$$J(t) = \frac{1}{2}e(t)^2 \tag{4.4.10}$$

$$\delta(t) = \xi_t e(t)\frac{dy(t)}{du(t)} \tag{4.4.11}$$

where e(t) = r(t) - y(t) and the binary factor was introduced to account for the constraints on the input u(t). Defining $\hat{e}(t) = e(t)[dy(t)/du(t)]$, for the system described in (4.4.1) and (4.4.2), at instant t the factor ξ_t is expressed as

$$\xi_t = \begin{cases} 0, & \text{if} \begin{cases} \hat{e}(t) > 0 \quad \text{and} \quad u(t-1) = u_M \\ \qquad\qquad \text{or} \\ \hat{e}(t) < 0 \quad \text{and} \quad u(t-1) = u_m \end{cases} \\ 1, & \text{otherwise} . \end{cases} \tag{4.4.12}$$

The role of ξ_t is to avoid mistraining the neuro-controller for references that cannot be tracked. The inclusion of ξ_t is equivalent to considering the existence of a limiter between the output of the neuro-controller and the input of the plant. When the output error results from the physical limitations expressed in (4.4.2), the limiter saturates and a zero-derivative ($\xi_t = 0$) is included in (4.4.11), inhibiting learning. On the other hand, when the reference can be physically followed, learning is performed in the usual way.

4.4.3.3 Emulator and Controller Scheme

Although in many practical cases the derivative in (4.4.11) can be easily estimated or replaced by + 1 or -1, that is, the signum of the derivative (since the algebraic sign of $\delta(t)$ is enough to specify the direction of the gradient of J), this is not the general case. In the emulator and controller scheme shown in Fig. 4.2.5, a PE is used to compute the sensitivity of the error function J with respect to the controller's output. Since PE is a multilayered neural network, the desired sensitivity can be easily calculated by using the backpropagation. Furthermore, the configuration in Fig. 4.2.5 is particularly useful when the inverse of the plant is ill-defined, i.e., the function $f_p(.)$ in (4.4.1) does not admit a true inverse. For updating the controller, the neuro-controller and PE are viewed as the variable and fixed parts of an m-input-single-output multilayered neural network, respectively. At the outset, the PE should be off-line trained with a data set sufficiently rich to allow plant identification, and then both the neuro-controller and the PE are on-line trained. In a sense, the PE performs system identification and, therefore, for rapidly changing system, it is reasonable to update the PE more often than the neuro-controller due to robustness.

4.4.4 Efficient On-Line Training

Multilayered neural networks and the backpropagation algorithm were originally developed for pattern classification problems, where the training patterns are static, the training procedure and the error function are straightforward, and real-time learning is not required. In control, training patterns for the neural network change with time, several training algorithms and error functions can be defined, and real-time learning is necessary.

Slow convergence is the most severe drawback of multilayered neural networks and seriously restricts neuro-control practical applications.

Several approaches have been proposed toward speeding up convergence in neuro-control. Among them, some common ideas are:

(1) Developing of efficient backpropagation algorithms.

(2) Embodiment of plant structural knowledge in the structure of the multilayered neural networks.

(3) Hybrid systems in which artificial neural networks are associated with control structures derived from non-neural techniques.

(4) Pre-learning and efficient initialization procedures.

Based on the distinction between sampling frequency and the frequency at which learning iterations are performed (learning frequency), new on-line training algorithms to reduce the neuro-controller training time are introduced here. In discrete-time control systems, the sampling period T is usually chosen via rules of thumb in order to have $2\pi/T$ much larger than the largest frequency involved in the continuous-time system. It is normally true that increasing the sampling frequency improves the system's performance, but the noticeable improvement rapidly reaches a plateau. In usual adaptive control systems, the adaptive elements are normally updated once in each sampling period in such a way that the sampling frequency and updating (or adaptation or learning) frequency may be used indistinctly. Ignoring processing time constraints, it seems that the actual training time can be reduced by increasing the sampling frequency.

However, in many practical applications the sampling frequency cannot or should not exceed some limit. For instance, in common industrial chemical plants, one is usually interested in processes involving large time constants. There is not much sense in using high sampling rates, since information redundancy would occur. In fact, very high sampling frequencies can modify the control system completely and increase its complexity. It may be necessary to consider fast sub-processes and transients that could be ignored otherwise. Another case in which the sampling frequency cannot be made arbitrarily high occurs in distributed control systems in which information is sent to and received from a control unit at intervals which are out of control of the unit.

Although the sampling period T sets the basic pace of the control system, in systems with iterative learning the frequency at which learning occurs can be thought of as a different time basis. For most practical cases,

T is much larger than T_L, the time spent in one learning iteration (updating of all network weights), and the ratio T/T_L tends to higher values as faster implementations of multilayered neural networks become available in hardware or software. Therefore, if appropriate plant input-output data is available and the only concern is time, many learning iterations can be performed during a sampling period and the normal and simplest approach of a single updating per period implies a waste of processing time. The problems are how to select appropriate training data and how to use such data and available time to perform meaningful training of the neuromorphic structures, that is, training that is likely to improve control performance. Novel training methods in which several learning iterations take place during each single sampling period are proposed in what follows [52].

4.4.4.1 Emulator's Training

Consider that at instant t + 1 the current plant output y(t + 1), the q + n - 1 previous values of y, and the p + n previous values of u are available in memory. Then the n pairs (x_E(t - i), y(t + 1 - i)), for i = 0, 1, ..., n - 1 can be used as patterns for training the PE at time t + 1. For y_1 (t + 1 - i) = φ_E (x_E(t - i)), one possibility is to minimize the following error function.

$$J_n(t) = \frac{1}{2}\left(\sum_{i=0}^{n-1} \lambda_i (y(t-i) - y_i(t-i))\right)^2 \qquad (4.4.13)$$

where $\{\lambda_1\}(1 \geq \lambda_0 \geq \lambda_1 \geq ... \geq \lambda_n - 1 \geq 0)$ is a non-increasing positive sequence whose role is to implement some "forgetting," emphasizing the most recent patterns.

Example 4.4.1 : Assume that y(10) was just read (and, therefore, y(11) is not available), p = 2, q = 3, and n = 3. Also assume that y(9), y(8), ... , y(5) and u(9), u(8), ... , u(5) are available in memory. Then the PE input vectors can be arranged as follows.

$$x_E(7) = [y(7), y(6), y(5), u(7), u(6), u(5)]^T$$

$$x_E(8) = [y(8), y(7), y(6), u(8), u(7), u(6)]^T$$

$$x_E(9) = [y(9), y(8), y(7), u(9), u(8), u(7)]^T.$$

These input vectors and values y(8), y(9) and y(10) constitute the training patterns for training at time t + 1 = 10. This procedure is illustrated in Fig. 4.4.1(a), where the notation $PE^{t,i}$ indicates the PE's state during the *t*th sampling interval, after the *i*th learning iteration, i = 0, 1, ..., n - 1. Equivalently, $\varphi_E^{t,i}$ (.) denotes the input-output mapping performed by $PE^{t,i}$. Clearly,

$$\varphi_E^{t,n}(.) = \varphi_E^{t+1,0}(.).$$

4.4.4.2 Controller's Training : The Direct Inverse Control Error Approach
Consider the direct inverse control configuration (Fig. 4.3.10) and assume that at instant t+1 the current output $y(t+1)$, the $q+n-1$ previous values of y and p + n previous values of u are all stored in memory. Then the n pairs $(x'_c(t-i), u(t-i))$, $i=0,\ldots,n-1$, for $x'_c(t) = [y(t+1),\ldots, y(t-q+1), u(t-1),\ldots, u(t-q)]^T$ can be used as patterns for training the neuro-controller at time t+1. Writing $u_1(t-i) = \varphi_c[x'_c(t-i)]$, for the error function

$$J^u(t) = 1/2 \sum_{i=0}^{n-1} \lambda_i (u(t-i) - u_i(t-i))^2, \tag{4.4.14}$$

the corresponding δ-term for the *i*th pattern becomes simply

$$\delta_{t,i} = \lambda_i (u(t-1) - u_i(t-i)). \tag{4.4.15}$$

Notice that the error function in (4.4.14) is not directly based on the plant output error. Hence, training of the neuro-controller does not improve control performance directly, unless learning has been carried out in such a way that good generalization through the control space can be expected. In

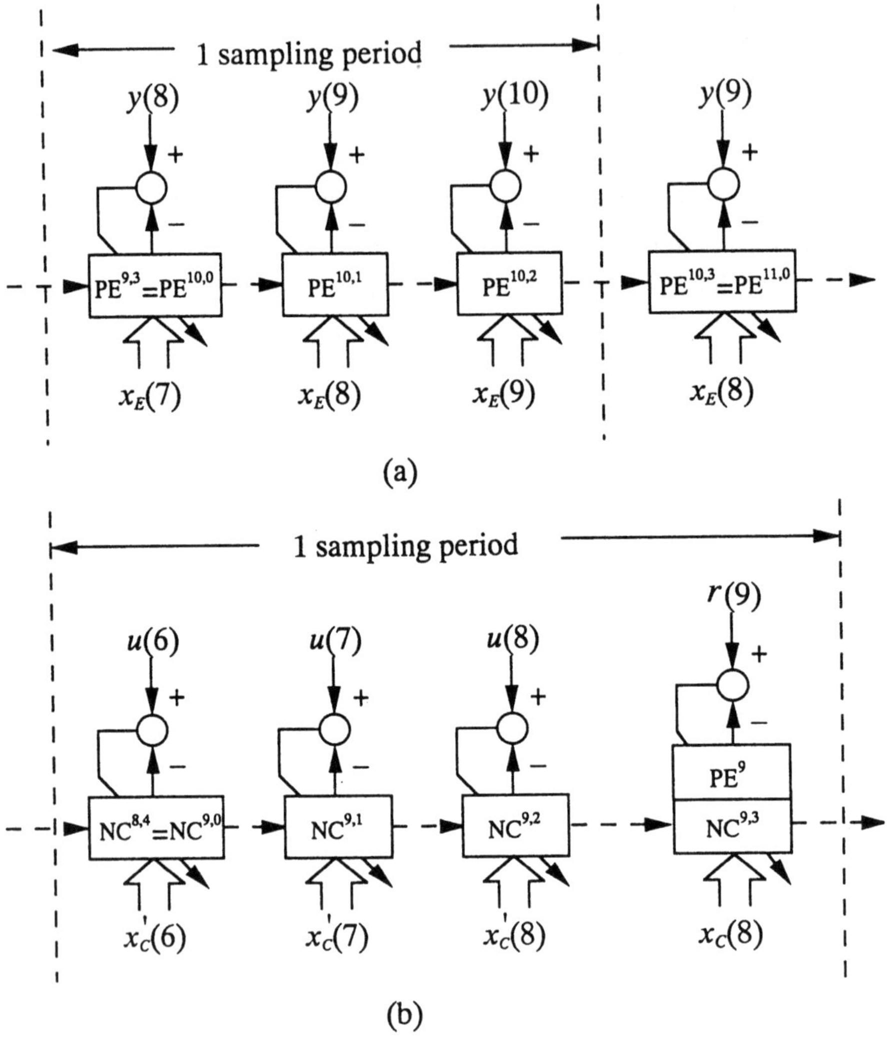

Fig. 4.4.1. Simple multiple training schemes (a) PE (Example 4.4.1). (b) Neuro-controller combining the direct inverse control and emulator and controller schemes (see Example 4.4.2) adopted from J. Tanomaru & S. Omatu [52] (© 1994 by IEEE inc.).

fact, the neuro-controller training based exclusively on the direct inverse control approach generally leads to bad results, and in practice, it has been observed that the output of neuro-controller tends to stick at some constant value, resulting in zero training error, but obviously poor control performance. This drawback, common to training methods based on the minimization of the inverse control error, can be overcome by combining such training methods with others that directly minimize the plant output error, as illustrated in the following example.

Example 4.4.2 : Assume that y(9) was just read, p = 3, and q = 2. Also assume that the values y(8), y(7), ..., y(5) and u(8), u(7), ..., u(3) are available in memory. Denoting the current mapping performed by the neuro-controller by $\varphi_C^{9,0}(.)$, (indicating t + 1 = 9, no learning yet), the plant input u(9) can be calculated from the relation $u(9) = \varphi_C^{9,0}[x_c(9)]$ where $x_c(9) = [r(10), y(9), y(8), u(8), u(7), u(6)]^T$.

For learning based on the direct inverse control error approach, the following vectors are then available:

$$x'_c(6) = [y(7), y(6), y(5), u(5), u(4), u(3)]^T$$
$$x'_c(7) = [y(8), y(7), y(6), u(6), u(5), u(4)]^T$$
$$x'_c(8) = [y(9), y(8), y(7), u(7), u(6), u(5)]^T.$$

These vectors and the input values u(6), u(7) and u(8) constitute three training patterns (input vector and desired output) available for training the neuro-controller at time t + 1 = 9. However, since this kind of training does not minimize the control error directly, in practice, it is necessary to combine this approach with one of the methods described in Sections 4.4.3.2 and 4.4.3.3. Figure 4.4.1(b) shows the situation in which multiple learning based on the direct inverse control and simple learning based on the control configuration are combined, resulting in four learning iterations per sampling period. The vector $x_c(8)$ is given by $[r(9), y(8), y(7), u(7), u(6), u(5)]^T$,the notation $NC^{t,i}$ denotes the neuro-controller's state during the *t*th sampling interval, after the *i*th learning iteration (the corresponding mapping is given by $\varphi_C^{t,i}(.)$) and the PE is considered perfectly trained for the sake of simplicity.

4.4.4.3 Controller's Training : The Predicted Output Error Approach

A more complex approach for multiple training of the neuro-controller can be derived from the emulator and controller configuration of Section 4.4.3.3. Assume that n reference values r (t + 1 - i), i = 0, 1, ..., n - 1, are also available at instant t + 1, in addition to n + p values of y, including y(t + 1), and n + q previous values of u. From (4.4.6) and (4.4.7), this is equivalent to having n input vectors x_c (t - i) in memory. According to (4.4.5), at instant t - i, the control input u(t - i) was generated by

$$u(t-i) = \varphi_C^{t-i,0}[x_c(t-i)]. \tag{4.4.16}$$

However, the neuro-controller has been updated several times since the time the vector x_c(t - i) was stored until the present instant t + 1. Hence, at time t + 1, the stored input vector x_c(t - i) would yield the virtual plant input

$$u^*(t-i) = \varphi_C^{t+1,0}[x_c(t-i)]. \tag{4.4.17}$$

This means that, although the most recent data x_c(t), with corresponding y (t + 1) and r(t + 1), can be directly used for training by one of the adaptive control configurations, the same does not happen with the past values $x_c(t-i)$, for $i = 1,\ldots, n-1$. For those cases, however, the corresponding plant responses can be predicted from the emulator, as shown in Fig. 4.4.2 (a), by

$$y^*(t+1-i) = \varphi_E^{t+1}[x_E^*(t-i)] \tag{4.4.18}$$

where the second superscript index of φ_E(.) was omitted for simplicity, and the vector x_E^* (t - i) is given by

$$x_E^*(t-i) = [y(t-i),\ldots,y(t-p+1-i),$$
$$u^*(t-i),\ldots,u^*(t-q-i)]^T. \tag{4.4.19}$$

The neuro-controller training can be understood by considering the neuro-controller and the PE as a single multilayered neural network ; at time t + 1, for each input vector x_c (t - i), i = 1, ..., n - 1, there corresponds a predicted error r(t + 1 - i) - y*(t + 1 - i), and training is performed in order to minimize some norm of the predicted output errors. A possible error function would be

$$J_n(t) = \frac{1}{2}\left(\sum_{i=0}^{n-1} \lambda_i (r(t-i) - y^*(t-i))\right)^2 . \tag{4.4.20}$$

In the training configuration shown in Fig. 4.4.2 (a), the neuro-controller is trained based on the error between the reference and the output of the PE, and not the error between the reference and the plant output, as in the direct adaptive control configuration in Fig. 4.4.2 (b). The previous values of the plant output y(t) as still necessary for the input vectors to the neuro-controller and PE.

<u>Example 4.4.3</u> : Assume that y(9) was just read, p = 3, and q = 2. Also assume that y(8), y(7), ..., 7(5) and u(8), u(7), ..., u(3) are available in memory, as well as r(9), r(8), and r(7). Therefore, the neuro-controller input vectors used to compute the three most recent control input values are available as follows:

$$x_c(6) = [y(7), y(6), y(5), u(5), u(4), u(3)]^T$$
$$x_c(7) = [y(8), y(7), y(6), u(6), u(5), u(4)]^T$$
$$x_c(8) = [y(9), y(8), y(7), u(7), u(6), u(5)]^T.$$

At time t + 1 = 9, a possible training procedure is shown in Fig. 4.4.2 (b), where three learning iterations take place during a sampling period. The predicted plant output values are computed following the scheme in Fig. 4.4.2 (a), enabling the utilization of $x_c(6)$ and $x_c(7)$ at time t + 1 = 9, whereas the most recent vector $x_c(8)$ is used in the conventional way.

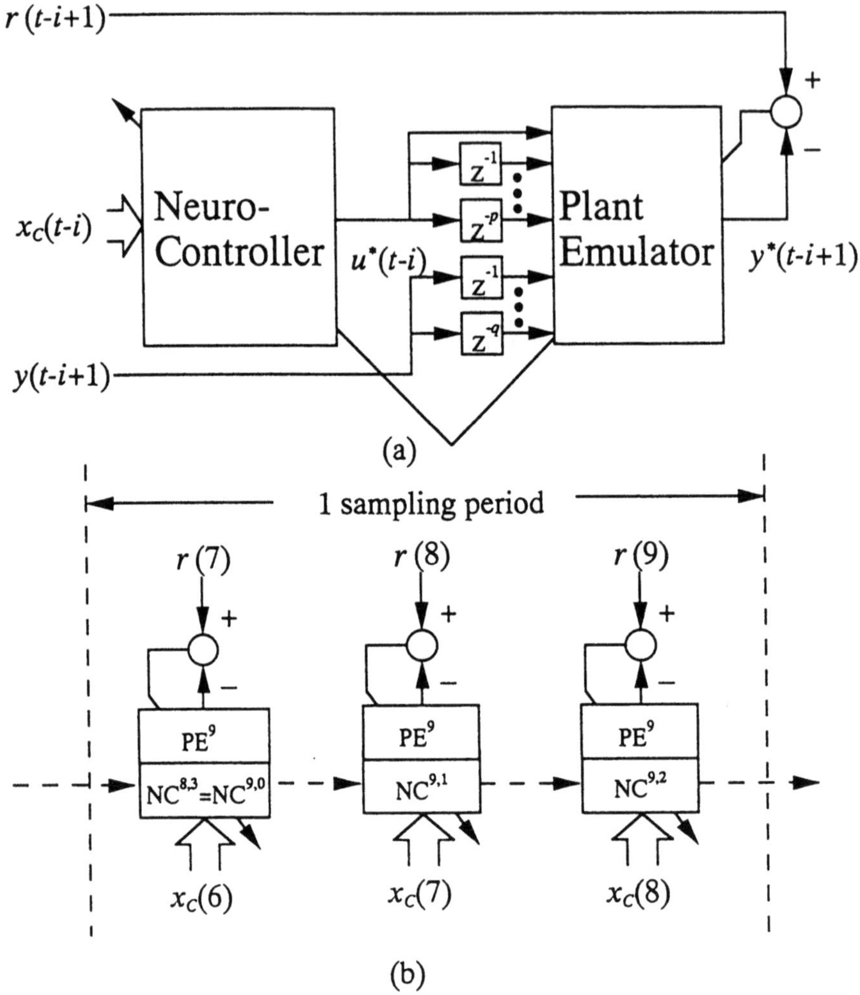

Fig. 4.4.2. Predicted output error approach. (a) Training scheme using errors between the reference and the output of the PE. (b) Multiple training of the neuro-controller by the predicted output error approach (see Example 4.4.3) adopted from J. Tanomaru &S. Omatu [52] (© 1994 by IEEE inc.).

4.4.5 Training Algorithms

In the following, the on-line training methods proposed in the previous section are written as algorithms in a sort of pseudo-code, making real implementation somewhat straightforward. The backpropagation algorithm is assumed to be known and is briefly represented by a routine call of the type

$$\mathbf{BP}(\phi, x, n, o)$$

where **BP** stands for the routine's name, ϕ denotes the network to be updated, x is the input vector, n the corresponding target (desired) output, and o is the actual output.

4.4.5.1 Multilearning Emulator

Assume that at instant t the vectors $x_{E,i} = [y(t-1-i), ..., y(t-q-i), u(t-1-i), ..., u(t-1-i-p]^T$, for $i = 0, ..., n-1$, are stored in memory. Clearly, the most recent data corresponds to $i = 0$, whereas higher values for i correspond to older data. Starting from the instant t, the algorithm for multiple-training of the emulator can be summarized by

Step 1. **READ** y(t)

Step 2. {emulator's training}

$i \leftarrow n-1$

REPEAT

$y_{1,i} \leftarrow \varphi_E(x_{E,i})$

$\mathbf{BP}(\varphi_E, x_{E,i}, \lambda_i y(t-i), \lambda_i, y_{1,i})$

$i \leftarrow i-1$

UNTIL $(i < 0)$

Step 3. { control input generation }

$x_c \leftarrow [r(t+1), y(t), ..., y(t+1-q), u(t-1),...,u(t-p)]^T$

or

$[r(t+1), r(t),...,r(t+1-q), u(t-1),..., u(t-p)]^T$

$u(t) \leftarrow \varphi_C(x_c)$

Step 4. **APPLY** u(t) to the plant and **WAIT** a T

Step 5. { data shifting }

$i \leftarrow n-1$

REPEAT

$x_{E,i} \leftarrow x_{E,i-1}$

$i \leftarrow i - 1$

UNTIL (i = 0)

Step 6. {most recent data vector)

$x_{E,0} \leftarrow [y(t), y(t-1), ..., y(t+1-q), u(t), ..., u(t-p)]^T$

Step 7. $t \leftarrow t + 1$

Step 8. **GO TO** Step 1.

4.4.5.2 *Multilearning Controller: The Direct Inverse Control Error Approach*

Assume that at instant t the vectors $x'_{c,i}=[y(t-i),..,y(t-q-i),u(t-2.-1),..,u(t-1-p-i)^T$, for i = 1, ..., n - 1, are available in memory. Starting from the instant t, the algorithm for the multilearning controller becomes

Step 1. **READ** y(t)

Step 2. { most recent data vector }

$x'_{c,0} \leftarrow [y(t),...., y(t-q), u(t-2), ..., u(t-p-1)]^T$

Step 3. { controller training }

$i \leftarrow n - 1$

REPEAT

$u_{1,i} \leftarrow \varphi_C(x'_{c,i})$

BP($\varphi_C, x'_{c,i}, \lambda_i u(t-1-i), \lambda_i u_{1,i}$)

$i \leftarrow n - 1$

UNTIL (i < 0)

Step 4. { control input generation }

$x_c = [r(t+1), y(t), ..., y(t+1-q), u(t-1),..., u(t-p)]^T$

or

$x_c=[r(t+1), r(t), ..., r(t+1-q), u(t-1),..., u(t-p)]^T$

$u(t) \leftarrow \varphi_C(x_c)$

Step 5. **APPLY** u(t) to the plant and **WAIT** a T

Step 6. { data shifting }

$i \leftarrow n-1$

REPEAT

$x'_{c,i} \leftarrow x'_{c,i-1}$

$i \leftarrow i-1$

UNTIL (i=0);

Step 7. $t \leftarrow t+1$

Step 8. **GO TO** Step 1.

4.4.5.3 Multilearning Controller : The Predicted Output Error Approach

Assume that the data corresponding the n + q + 1 vector $x_{c,i} = [r(t-i), y(t-1-i), \ldots, y(t-q-i), u(t-2-i), \ldots, u(t-p-1-i)]^T$, or alternatively, $x_{c,i} = [r(t-i), r(t-1-i), \ldots, r(t-q-i), u(t-2-i), \ldots, u(t-p-1-i)]^T$, are available in memory at instant t. Denoting by $(\varphi_C + \varphi_E)$ the neural networks formed by the association between the neuro-controller and the PE, starting from instant t, the algorithm for multiple-learning of the controller becomes

Step 1. **READ** y(t)

Step 2. { controller's training via the predicted error approach }

$i \leftarrow n-1$

REPEAT

$j \leftarrow 0$

REPEAT

$u^*_j \leftarrow \varphi_C(x_{c,i+j})$

$j \leftarrow j+1$

UNTIL (j > p)

{virtual input vector for the emulator}

$x_E \leftarrow [y(t-1-i), \ldots, y(t-q-i), u^*_0, u^*_1, \ldots, u^*_p]^T$

{virtual output}

$y^* \leftarrow \varphi_E(x_E)$

BP$(\varphi_C + \varphi_E, x_{c,i}, \lambda_i r(t-i), \lambda_i y^*)$

$i \leftarrow i-1$

UNTIL (i=0)

Step 3. {conventional training using most recent data}

BP($\varphi_C + \varphi_E$, $x_{c,i}$, $\lambda_i r(t)$, $\lambda_i y(t)$)

Step 4. {data shifting}

$i \leftarrow n + q + 1$

REPEAT

$x_{c,i} \leftarrow x_{c,i-1}$

$i \leftarrow i - 1$

UNTIL (i=0)

Step 5. {control input generation}

$x_{c,0} = [r(t+1), y(t), ..., y(t+1-q), u(t-1), ..., u(t-p)]^T$

or

$[r(t+1), r(t), ..., r(t+1-q), u(t-1), ..., u(t-p)]^T$

$u(t) \leftarrow \varphi_C(x_{c,0})$

Step 6. **APPLY** u(t) to the plant and **WAIT** a T

Step 7. $t \leftarrow t + 1$

Step 8. **GO TO** Step 1.

4.4.6 Evaluation of the Training Algorithms Through Simulation

In order to evaluate the performance of the proposed control system and training methods, results from simulation of a simple plant model are presented. Computer simulation is especially appealing when dealing with on-line training of neuro-control system, since very time-consuming experiments are often necessary. Consider the following continuous-time temperature control system

$$\frac{dy(\tau)}{d\tau} = \frac{f(\tau)}{C} + \frac{Y_0 - y(\tau)}{RC} \tag{4.4.21}$$

where τ denotes time, $y(\tau)$ is the output temperature in °C, $f(\tau)$ is the heat flowing inward the system, Y_0 is the room temperature (constant, for simplicity), C denotes the system thermal capacity, and R is the thermal resistance between the system borders and surroundings. Assuming that R and C are essentially constant, obtaining the pulse transfer function for the system in (4.4.21) by the step response criterion results in the discrete-time system.

$$y(\tau+1) = a(T)\, y(\tau) + b(T)\, u(\tau) \tag{4.4.22}$$

where τ is the discrete-time index, T is a sampling interval, $u(\tau)$ and $y(\tau)$ denote the system input and output, respectively. Denoting by α and β some constant values depending on R and C, the remaining parameters can be expressed by

$$a(T) = e^{-\alpha T} \qquad \text{and} \qquad b(T) = \frac{\beta}{\alpha}(1-e^{-\alpha T}). \tag{4.4.23}$$

For the simulation results presented below, the system described by (4.4.21) - (4.4.23) was modified to include a saturating nonlinearity so that the output temperature cannot exceed some limitation. The simulated control plant is described by

$$y(t+1) = a(T)\, y(t) + \frac{b(T)}{1+e^{0.5y(t)-\gamma}} u(t) + [1 - a(T)]Y_0. \tag{4.4.24}$$

The parameters for simulation are $\alpha = 1.00151\text{x}10^{-4}$, $\beta = 8.67973\text{x}10^{-3}$, $\gamma =$ 40.0, and $Y_0 = 25.0°C$, which were obtained from a real water bath plant (see Chapter 5). The plant input u(t) was limited between 0 and 5, and it is also assumed that the sampling period is limited by

$$T \geq 10 \text{ sec.} \tag{4.4.25}$$

With the chosen parameters, the simulated system is equivalent to a SISO temperature control system of a water bath that exhibits linear behaviour up to about 70°C and then becomes nonlinear and saturates at about 80°C. Comparing (4.4.24) with (4.4.1), it is clear that $\bar{p} = 0$ and $\bar{q} = 1$.

Simple three-layered neural networks with 10 to 20 hidden sigmoidal neurons were chosen for the PE and neuro-controller, and convergence was obtained for several pairs (p, q), for p ranging from 0 to 2, and q from 1 to 3. The graphs presented in the following correspond to p = 0 and q = 1 (perfect matching). The networks were updated according to the general rule

$$\Delta w(t) = w(t+1) - w(t)$$

$$\Delta w(t) = -\eta \frac{\partial J}{\partial w(t)} + \alpha \Delta w(t-1) \tag{4.4.26}$$

where t denotes learning iterations, w is a connection weight, J is the error function to be minimized, $\eta > 0$ is the learning rate, and $\alpha \geq 0$ is the momentum term.

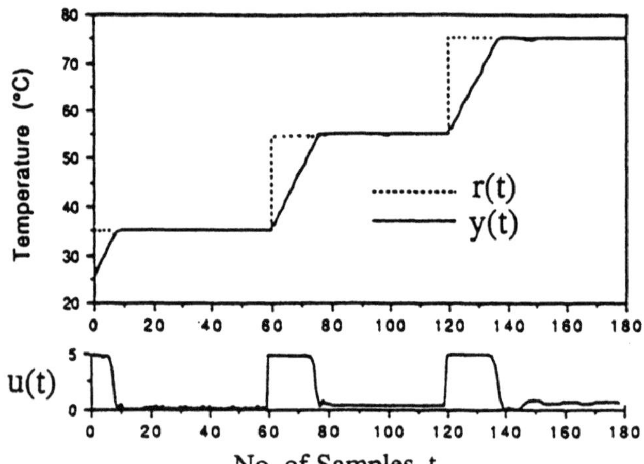

Fig. 4.4.3. Simulation result of the water bath system after learning.adopted from J. Tanomaru & S. Omatu [52] (© 1994 by IEEE inc.).

Before the outset, the PE is roughly off-line trained, whereas the weights of the neuro-controller are randomly initialized. From the initial condition $y(0) = Y_0$, the target is to follow a control reference set to 35 °C for $0 \leq tT \leq 30$ min., 55 °C for 30 min. $< tT \leq 60$ min., and 75 °C for 60 min. $< tT \leq 90$ min. Each simulation cycle, for tT ranging from 0 to 90 min., is called a trial. After a trial, the weights are conserved and a new trial starts. Figure 4.4.3 shows the reference, and the plant input and output after good convergence was achieved for T = 30sec.

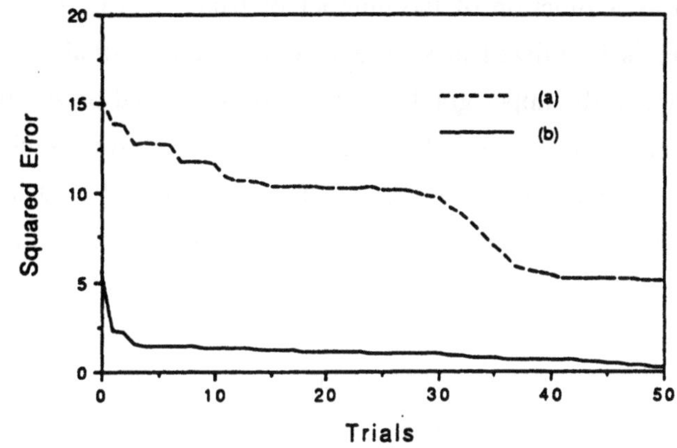

Fig. 4.4.4. Neuro-controller randomly initialized for T = 30 sec. (a) Only conventional on-line training. (b) Conventional training plus 5 learning iterations/period based on the predicted output error approach. (c) Conventional training plus 10 learning iterations/ period adopted from J. Tanomaru & S. Omatu [52] (© 1994 by IEEE inc.).

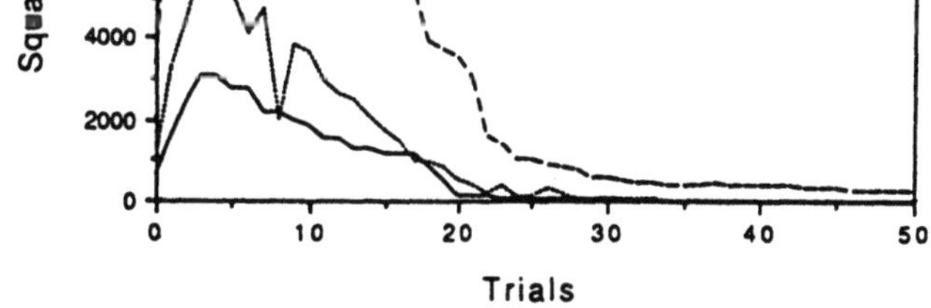

Fig. 4.4.5. Neuro-controller randomly initialized performing one learning iteration per sampling period. (a) T = 30 sec. (b) T = 15 sec. (c) T = 10 sec. adopted from J. Tanomaru & S. Omatu [52] (© 1994 by IEEE inc.).

The performance of neuro-controller carrying out a single learning iteration per sampling period is compared with the multiple-learning-iteration case in Figs. 4.4.4-4.4.6. The graphs show the total squared error per trial as a function of the number of trials. In Fig. 4.4.4 the neuro-controller is initialized in such a way to result in small error from the beginning. In the upper graphs, training is performed once a period using the indirect adaptive control configuration. In the lower graph, the neuro-controller is updated 10 additional times per period by using the direct inverse control error approach. In both cases the sampling period is T = 30 sec.

The proposed method improved performance as expected, since the neuro-controller is assumed to be relatively well trained from the outset and generalization is, thus, reliable. This fact suggests that this training method can be used for small adjustments of the neuro-controller near a good operating point, i.e., for fine tuning. The same is not true when the neuro-controller is in a naive state. In fact, experience indicates that when the weights of the neuro-controller are such that the plant output error is large, training exclusively based on the inverse control approach often leads to situations in which the output of the neuro-controller sticks at some value, resulting in poor performance.

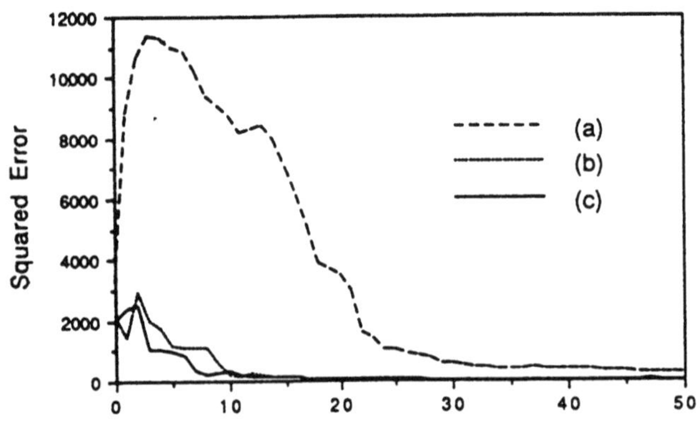

Fig. 4.4.6 Neural controller randomly initialized for T = 30 sec. (a) Only conventional on-line training. (b) Conventional training plus 5 learning iterations/period based on the predicted output error approach. (c) Conventional training plus 10 learning iterations/ period adopted from J. Tanomaru & S. Omatu [52] (© 1994 by IEEE inc.).

The performance of a randomly initialized neuro-controller carrying out one learning iteration per period via the indirect adaptive control configuration is shown in Figs. 4.4.5(a) and 4.4.6(a) for T = 30 sec. Since each simulation trial is equivalent to 90 min. of real time, 50 trials would require 75 hours or more than three days, and that is one of the reasons why most of the results presented here involve simulations rather than real experiments.

In the lower graphs in Fig. 4.4.5, the sampling period was changed to 15 sec. (Fig. 4.4.5 (b)) and 10sec. (Fig.4.4.5 (c)), and an expected performance improvement was observed. On the other hand, in the remaining graphs in Fig. 4.4.6, the sampling period was fixed in 30 sec., and 5 or 10 additional learning iterations per period were included, based on the proposed predicted output error approach (Figs. 4.4.4 (b) and 4.4.6 (c), respectively). The inclusion of only a few learning iterations per sampling period resulted in sharp convergence speed-up, greatly reducing the total error.

The time spent for updating the neuromorphic structure is roughly proportional to the number of learning iterations, and basically depends on the structure and number of weights of the networks being considered. In the simulation results presented here, 11 learning iterations of a three-layered neuro-controller with 80 weights took about 11 x T_L = 390 msecs. in a personal computer, far less than the sampling period in many practical control applications.

As mentioned above, the results in Figs. 4.4.3-4.4.6 were obtained for $\overline{p}$ = p = 0 and $\overline{q}$ = q = 1. What happens when one does not have accurate estimates of the order of the plant is a question that arises naturally. In the simulations carried out, training has succeeded for several combinations of different p and q values, but much caution is recommended when generalizing results. Small changes in the initial conditions, references, structures of the neural nets, neuron activation functions, and so forth too often produce dramatic consequences in the performance results. Various combinations for p and q were tested under different conditions, and some results are illustrated in Fig. 4.4.7. The perfect matching case is shown in Fig. 4.4.7 (a), whereas Fig. 4.4.7 (b) and (c) correspond to mismatching cases for p = 0, q = 3 and p = 2, q= 2, respectively. Although in some cases, as in Fig. 4.4.7 (c), convergence could not be obtained, good results were achieved for several pairs (p, q) near the true values $\bar{p}$ and $\bar{q}$.

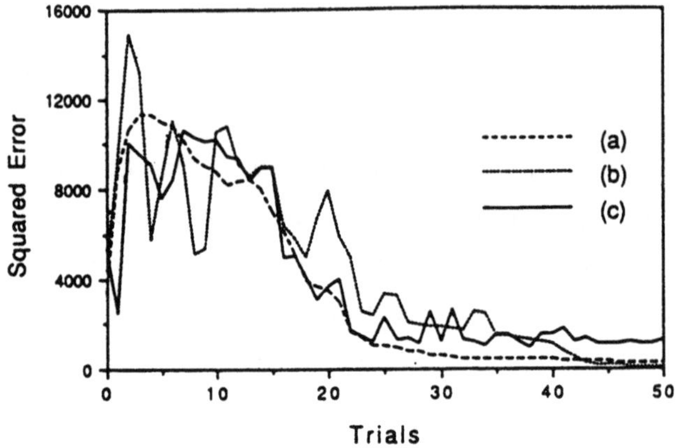

Fig. 4.4.7. Effects of mismatching between the estimates and the optimal order of the plant model. (a) Perfect matching for $p=\bar{p}=0$, $q=\bar{q}=1$ (b) Mismatching with convergence for p=0, q=3. (c) Mismatching without convergence for p=q=2 adopted from J. Tanomaru & S. Omatu [52] (© 1994 by IEEE inc.).

4.5 Parallel Control Scheme

A parallel neuro-control architecture is shown as in Fig. 4.5.1 where a PID controller is used as the conventional controller. The feedback-error learning method proposed by Kawato *et al.* [40], [41] as described in Section 4.2.1 is an example of this type of neuro-control scheme. A more detailed explanation of this scheme will be described in the next section.

This parallel type neuro-controller is used to adjust the control input u_1 to the plant, which is the output of a conventional controller, such that the plant output y could follow a desired reference signal r as precisely as possible. The function of the parallel type neuro-controller is to adjust a conventional control input u_1 if it cannot provide good results. For example, a well-trained cook at the restaurant could provide a delicious dinner for his diners, however, on each table there are pepper and salt to be added to suit the taste of each individual diner.

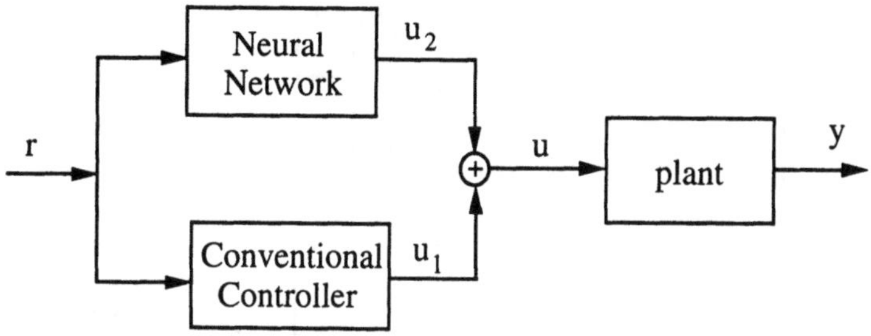

Fig. 4.5.1. A parallel neuro-control scheme.

As another example, control engineers design an excellent controller at the laboratory or factory which is given by u_1 but when it is set at the real working place in an industrial factory, the control engineers must adjust the control input level such that it is suitable for real production. These examples mean that even if the experts could produce good control strategies, it might need to be re-adjusted again to achieve better practical outcomes.

4.5.1 Learning Algorithm for Parallel Control Scheme

The general control configuration is illustrated in Fig. 4.5.2 for a parallel type neuro-controller. In the figure, NN1 is a neural network for plant emulator which provides an estimate $\hat{y}$ of the plant output y. The control signal u_2 which is an output by the neural network NN2 is used for correcting the control signal u_1 produced by the conventional controller such that the error between the reference signal and plant output is minimized. Thus, NN1 is used to emulate the system Jacobian which is required to obtain the equivalent error at the output of NN2.

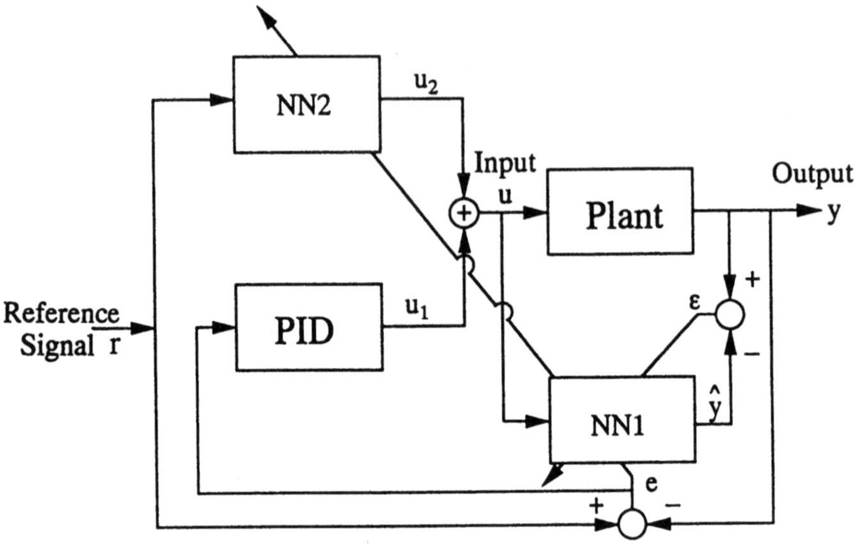

Fig. 4.5.2. Parallel type neuro-controller with emulator.

Let us denote the error between the reference signal r and the actual plant output y by e. The objective is to train NN2 such that a squared error can be minimized. Let us denote the squared error by E such that

$$E = \frac{1}{2} r - y^2. \qquad (4.5.1)$$

The output of NN2 is denoted by u_2 as shown in Fig. 4.5.2. We assume a three-layered neural network whose input is $r_i(t)$ and output at the output layer is $O_k(t)$ where i = 1 and k = 1 for simplicity. Thus, we will represent $r_i(t)$ and $O_k(t)$ by r(t) and O(t), respectively. Furthermore, assume that the plant has one time-delay. Then we have

$$y(t+1) = f_p(u(t)), \ u(t) = u_1(t)+u_2(t), \ u_2(t) = O(t)\cdot \qquad (4.5.2)$$

To derive the learning algorithm, we use the steepest descent (gradient) method to minimize E. Thus, we have

$$\Delta w_j(t+1) = -\eta \frac{\partial E}{\partial w_j} + \alpha \Delta w_j(t) \tag{4.5.3}$$

$$\Delta w_i(t+1) = -\eta \frac{\partial E}{\partial w_i} + \alpha \Delta w_i(t) \tag{4.5.4}$$

where η and α are the learning rate and momentum term, respectively. w_j and w_i denote connection weights from neuron j at the hidden layer to the output neuron at output layer and from the input neuron at input layer to the neuron i at the hidden layer, respectively, and Δw is the increment of the connection weight given by

$$\Delta w = w(\text{new}) - w(\text{old}). \tag{4.5.5}$$

Denoting $\delta.$ by

$$\delta. = \frac{\partial E}{\partial net.} \tag{4.5.6}$$

where

$$net. = \sum_j w_j O_j + \theta. \tag{4.5.7}$$

then we have

$$\frac{\partial net.}{\partial w_j} = \frac{\partial E}{\partial net.} \frac{\partial net.}{\partial w_j} = -\delta. O_j . \tag{4.5.8}$$

Using sigmoid (logistic activation) functions the input/output relation of the neuron is

$$O.(t) = f(net.)$$

$$f(x) = \frac{1}{1+e^{-x}}$$

and from the input/output relation of the plant we have

$$y(t+1) = f_p(u(t)),\ u(t) = u_1(t)+u_2(t),\ u_2(t)=O(t). \tag{4.5.9}$$

Therefore, we have

$$\begin{aligned}\delta_. &= -\frac{\partial E}{\partial y(t+1)}\frac{\partial y(t+1)}{\partial u(t)}\frac{\partial u(t)}{\partial net_.} \\ &= (r(t+1) - y(t+1))\frac{\partial y(t+1)}{\partial u(t)}\left(\frac{\partial u_1(t)}{\partial net_.} + \frac{\partial u_2(t)}{\partial net_.}\right) \\ &= (r(t+1) - y(t+1))\frac{\partial y(t+1)}{\partial u(t)} u_2(t)(1-u_2(t))\end{aligned} \tag{4.5.10}$$

where $\partial u_1/\partial net_. = 0$ was used.

If we take the activation function at the output layer as an identity mapping, that is, $f(x) = x$, then we have

$$\delta^I_. = (r(t+1) - y(t+1))\frac{dy(t+1)}{du(t)} \tag{4.5.11}$$

where $\delta^I_.$ denotes $\delta_.$ for the identity mapping.
Thus, we have

$$\Delta w_j(t+1) = -\eta\delta_. O_j + \alpha\Delta w_j(t) \tag{4.5.12}$$

or

$$\Delta w_j(t+1) = -\eta\delta^I_. O_j + \alpha\Delta w_j(t). \tag{4.5.13}$$

For unit i at the hidden layer, we have

$$O_i = f(net_i)$$

$$net_i = w_i r + \theta_i$$

Defining δ_i as

$$\delta_i = -\frac{\partial E}{\partial net_i}, \tag{4.5.14}$$

we obtain

$$\begin{aligned}\delta_i &= -\frac{\partial E}{\partial net_i} = -\frac{\partial E}{\partial net.}\frac{\partial net.}{\partial O_i}\frac{\partial O_i}{\partial net_i} \\ &= \delta.w_i f'(net_i) = \delta.w_i O_i(1-O_i)\end{aligned} \tag{4.5.15}$$

and from chain rule we have

$$\frac{\partial E}{\partial w_i} = \frac{\partial E}{\partial net_i}\frac{\partial net_i}{\partial w_i} = -\delta_i r. \tag{4.5.16}$$

Thus, we have a learning rule at the hidden nodes as follows:

$$\Delta w_i(t+1) = \eta\delta_i r + \alpha\Delta w_i(t). \tag{4.5.17}$$

Therefore, the rule for a parallel type neuro-control scheme is given by the following steps.

Step 1. Set the initial values of w_j, w_i, θ_j, θ_i, η, and α.
Begin $t = 0$ and go to step 2.

Step 2. Apply the desired outputs r(t) as the input to the neural network.
Then compute $\delta.$ or $\delta^I_.$ by the formula

$$\delta. = u(t)\,(1-u(t))\;e(t+1)f_p'(u(t))$$

or

$$\delta^I_. = e(t+1)\,f'_p(u(t)).$$

where $e(t+1) = r(t+1) - y(t+1)$.

Step 3. Change the connection weights by

$$\Delta w_j(t+1) = \eta \delta . O_j \quad + \quad \alpha \, \Delta w_j(t).$$

or

$$\Delta w_j(t+1) = \eta \delta^I_{.} O_j \quad + \quad \alpha \, \Delta w_j(t).$$

Step 4. Calculate δ_i or δ^I_i by

$$\delta_i = w_i \, \delta . \, O_i \, (1 - O_i).$$

or

$$\delta^I_i \; = \; w_i \, \delta.$$

Step 5. Change the connection weights by

$$\Delta w_i(t+1) = \eta \delta_i r(t) \quad + \quad \alpha \, \Delta w_i(t)$$

or

$$\Delta w_i(t+1) = \eta \delta^I_i r(t) \quad + \quad \alpha \, \Delta w_i(t).$$

Step 6. $t \rightarrow t + 1$ and go to Step 2.

Note that as in the case of a series type neuro-controller, the learning rule needs the plant Jacobian $f'_p(u(t)) = \partial y(t+1)/\partial u(t)$. This can be approximated by numerical difference or using the emulator network NN1 described in Fig. 4.5.2.

4.6 Feedback-Error Learning Algorithm

This algorithm has been proposed by M. Kawato *et al.* [40], [41] which is a special case of the parallel type neuro-control scheme. In the original feedback-error learning scheme as shown in Fig. 4.2.4, the feedback-error which is the output of the feedback controller is fed back through the neural network layers over thousands of learning cycles until convergence is achieved. The neural network then takes control as the dominant controller for the plant eliminating the effect of the conventional controller. The schematic diagram of the feedback-error learning control was shown in Fig. 4.2.4.

In this section we describe a modified version of the feedback error learning scheme similar to the configuration as shown in Fig. 4.5.2. For learning, we minimize the error function E given by (4.5.1). NN2 is assumed to be a three-layered neural network and the activation function at the output layer is an identity mapping, that is, f(x) = x. Using the steepest descent (gradient) algorithm, we have at the output layer

$$\Delta w_{kj}(t+1) = -\eta \frac{\partial E}{\partial w_{kj}} + \alpha \Delta w_{kj}(t) \tag{4.6.1}$$

and at the hidden layer

$$\Delta w_{ji}(t+1) = -\eta \frac{\partial E}{\partial w_{ji}} + \alpha \Delta w_{ji}(t). \tag{4.6.2}$$

Using thc chain rulc, we have

$$\frac{\partial E}{\partial w_{kj}} = \frac{\partial E}{\partial net_k} \frac{\partial net_k}{\partial w_{kj}} \tag{4.6.3}$$

$$\frac{\partial E}{\partial net_k} = \frac{\partial E}{\partial y} \frac{\partial y}{\partial u} \left(\frac{\partial u}{\partial u_2} \frac{\partial u_2}{\partial net_k} + \frac{\partial u}{\partial u_1} \frac{\partial u_1}{\partial net_k} \right)$$

$$= -(r-y) \frac{\partial y}{\partial u} \frac{\partial u_2}{\partial net_k} = -(r-y) \frac{\partial y}{\partial u} \tag{4.6.4}$$

where u_1 and u_2 are the control inputs given by the conventional controller and the neuro-controller, respectively. Here,

$$u_2 = net_k = \sum_j w_{kj} O_j + \theta_k$$

and

$$u = u_1 + u_2. \tag{4.6.5}$$

Thus, if we define δ_k and δ_j by

$$\delta_k = -\frac{\partial E}{\partial net_k}$$
$$\delta_j = -\frac{\partial E}{\partial net_j}, \tag{4.6.6}$$

we have

$$\Delta w_{kj}(t+1) = \eta \delta_k O_j + \alpha \Delta w_{kj}(t) \tag{4.6.7}$$

$$\delta_k = e\frac{\partial y}{\partial u} = e\left|\frac{\partial y}{\partial u}\right| \text{sign}\left(\frac{\partial y}{\partial u}\right) \tag{4.6.8}$$

$$e = r - y \tag{4.6.9}$$

$$\Delta w_{ji}(t+1) = \eta \delta_j O_i + \alpha \Delta w_{ji}(t) \tag{4.6.10}$$

$$\delta_j = w_{kj} \delta_k O_j (1 - O_j) \tag{4.6.11}$$

where sign (x) = 1 for $x \geq 0$ and = -1 for $x < 0$ and $O_j = f(net_j)$ is the output of the neuron j at the hidden layer. This algorithm also requires the plant Jacobian $\frac{\partial y}{\partial u}$. One way is to use the emulator described by Fig. 4.5.2 and another way is to find only the sign of the Jacobian and select the learning rate η by including the absolute value $\left|\frac{\partial y}{\partial u}\right|$.

The feedback-error learning algorithm is based on motion control of the brain. One biological example of this learning scheme has been given in Section 4.2.1. This kind of learning mechanism for our brain has been proven biologically in brain science [54]-[56]. For practical use, we can adopt a better modeling approach for the neural networks of NN1 and NN2. They can be modeled by using the architectures described by Figs. 4.3.4 and 4.3.6, or in order to speed up the modeling of the emulator, the structure of Fig. 4.3.7 is recommended from our experience.

4.7 Extension of the Parallel Type Neuro-Controller

Another extension of the parallel type neuro-controller is to apply the technique to nonlinear control problems. This technique has been proposed by Y. Iiguni *et al.* [13] and extended by H. Kinjo *et al.* [57]. The basic idea of this approach is to make a linear mathematical model as a first order approximation of a nonlinear dynamical system and to apply the neural network to improve the discrepancy between the output of the model and the actual output of the system. If a linear model is assumed, we can determine the optimal control law for the quadratic cost function, however, this is not the true control law since it is derived using a linearized mathematical model. In order to adjust the control input such that the quadratic cost function is minimized for the real plant, a neural network is introduced. This algorithm is described in the sections that follow.

4.7.1 Description of Control System

Consider a general plant described by

$$x(t+1) = A\,x(t) + A^{*}\,(x(t)) + bu(t) \tag{4.7.1}$$

$$y(t+1) = c\,x(t+1) \tag{4.7.2}$$

where u(t) and y(t) are scalar-valued functions denoting input and output of the system, respectively, $x(t) = [x_1(t), x_2(t), \dots, x_n(t)]^T$ is an n-dimensional state vector and A, b, c are system parameters and A^* denotes the nonlinear part of the system. This term provides the additional term when we fit the linear model as the plant model. The control problem is to find the optimal control solution which minimizes the cost function J given by

$$J = \sum_{t=0}^{\infty} (y^2(t+1) + \gamma u^2(t)) \tag{4.7.3}$$

where γ is a positive constant.

4.7.2 Linearized Control System

Figure 4.7.1 shows the optimal control system for an approximated linear system based on linear quadratic control theory. Here, we denote $\bar{A}$ and $\bar{b}$ as the approximated system parameters. Using the Luenberger observer, the state estimation algorithm is given by

$$
\begin{aligned}
\bar{x}^a(t+1) &= \bar{A}\bar{x}(t) + \bar{b}u(t) \\
\bar{y}^a(t+1) &= C\bar{x}^a(t+1) \\
\bar{x}(t+1) &= \bar{x}^a(t+1) + f(y(t+1) - \bar{y}^a(t+1))
\end{aligned}
\tag{4.7.4}
$$

where $\bar{x}^a$ and $\bar{x}$ denote *a priori* and *a posteriori* estimates of x, respectively and f is an observer gain which is specified by a conventional design rule of the observer. Using optimal control theory, the optimal control input u(t) for the approximated linear system with minimizing cost function J is given by

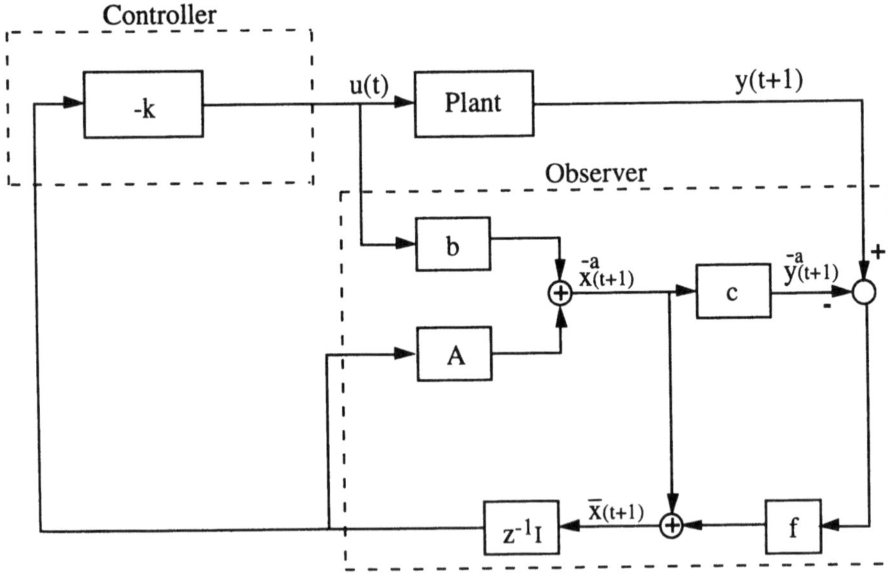

Fig. 4.7.1. Optimal control system for a linearized system where I denotes an identity operator, and z^{-1} is a time-delay operator.

$$u(t) = -k\,\bar{x}(t) \tag{4.7.5}$$

where k is a regulator gain row vector of n-dimension which is also obtained based on the matrix Riccati algorithm [57].

4.7.3 Control System with Neural Networks

Figure 4.7.2 shows a control system compensated with neural networks where neural networks are used to compensate the discrepancies for control input and estimate by the observer. Here, NN1 compensates the nonlinearity included in the system equation which cannot be estimated by a linear observer. NN2 is used to adjust the control input such that the system output approaches to zero.

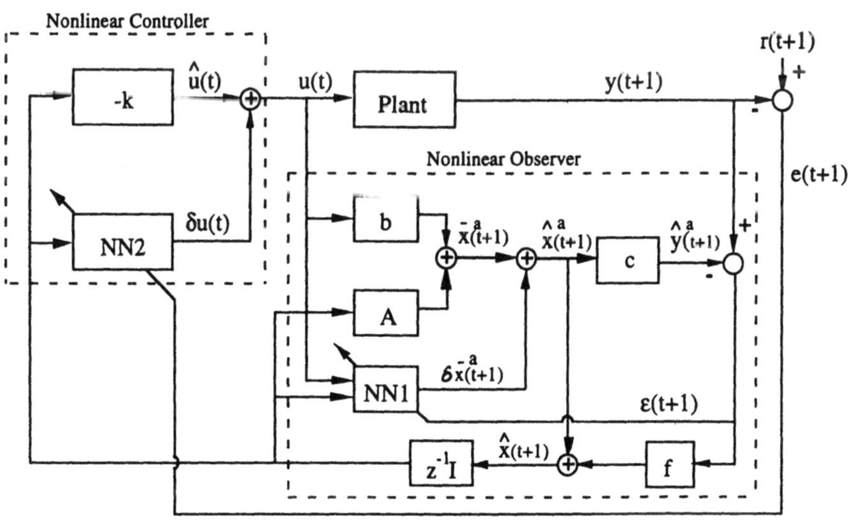

Fig. 4.7.2. Control systems using neural networks where r(t+1) denotes the desired signal and it is assumed here that r(t+1) = 0.

4.7.4 Nonlinear Observer by Neural Network

As shown in Fig. 4.7.2, the *a priori* estimate $\bar{x}^a(t+1)$ using the Luenberger observer is given by

$$\bar{x}^a(t+1) = \bar{A}\,\bar{x}(t) + bu(t) \tag{4.7.6}$$

where $\bar{x}(t)$ is the best estimate of x(t). However, $\bar{x}^a(t+1)$ does not include any nonlinear property of the state equation. Thus, we adjust the *a priori* estimate by $\delta\bar{x}^a(t+1)$ using the neural network NN1 in Fig. 4.7.2. Let us define $\hat{x}^a(t+1)$ by

$$\hat{x}^a(t+1) = \bar{x}^a(t+1) + \delta\bar{x}^a(t+1) \tag{4.7.7}$$

where $\delta\bar{x}^a(t+1)$ is the output of neural network NN1, which is trained such that the output error $\varepsilon(t+1)$ is minimized.

$$\begin{aligned} \varepsilon(t+1) &= y(t+1) - \hat{y}^a(t+1) \\ \hat{y}^a(t+1) &= c\,\hat{x}^a(t+1). \end{aligned} \tag{4.7.8}$$

Then the estimate $\hat{x}(t+1)$ of $x(t+1)$ is given by

$$\hat{x}(t+1) = \hat{x}^a(t+1) + f(y(t+1) - y^a(t+1)) \tag{4.7.9}$$

where the learning rule is based on the backpropagation which minimizes the squared error $\hat{E}$ given by

$$\hat{E} = \frac{1}{2}\varepsilon^2(t+1). \tag{4.7.10}$$

If we define $y^r(t+1) = \frac{1}{2}\varepsilon^2(t+1)$. by

$$y^r(t+1) = y(t+1) - \bar{y}^a(t+1) = c\big(x(t+1) - \bar{x}^a(t+1)\big) \tag{4.7.11}$$

and $\delta y(t+1)$ by

$$\delta y(t+1) = c\delta \bar{x}^a(t+1),$$

where

$$\bar{y}^a(t+1) = \bar{x}^a(t+1)$$

then $\hat{E}$ given by (4.7.10) becomes

$$\hat{E} = \frac{1}{2}(y^r(t+1) - \delta y(t+1))^2. \tag{4.7.12}$$

Thus, if we regard $y^r(t+1)$ as the desired signal and $\delta y(t+1)$ as the output part given by the neural network NN1, then the backpropagation method can be applied directly to minimize $\hat{E}$. Here, the output part of the neural network NN1 is $\delta y(t+1) = c\delta \bar{x}^a(t+1)$ and the inputs of NN1 is $\bar{x}(t)$ and u(t).

4.7.5 Nonlinear Controller by Neural Network

As shown in Fig. 4.7.2, the control input u(t) is given by

$$u(t) = \hat{u}(t) + \delta u(t) \tag{4.7.13}$$

$$\hat{u}(t) = -k\hat{x}(t) \tag{4.7.14}$$

where k is the same feedback gain obtained in Section 4.7.2 and $\delta u(t)$ is the output of the neural network NN2 which minimizes the cost function E given by

$$E = \frac{1}{2}\left(e^2(t+1) + wu^2(t)\right). \tag{4.7.15}$$

Thus, if we denote the connection weights at the output layer and the hidden layer by w_{kj} and w_{ji}, respectively, then we have

$$\Delta w_{kj}(t+1) = -\eta \frac{\partial E}{\partial w_{kj}} + \alpha \Delta w_{kj}(t)$$

$$\Delta w_{ji}(t+1) = -\eta \frac{\partial E}{\partial w_{ji}} + \alpha \Delta w_{ji}(t) \qquad (4.7.16)$$

where k=1 since number of neurons at the output layer is equal to one. Using chain rule, we have

$$\frac{\partial E}{\partial w_{kj}} = \frac{\partial E}{\partial net_k} \frac{\partial net_k}{\partial w_{kj}} \qquad (4.7.17)$$

where

$$net_k = \sum_j w_{kj} O_j. \qquad (4.7.18)$$

Again, by applying the chain rule to $\partial E/\partial net_k$, we have

$$\frac{\partial E}{\partial net_k} = \frac{\partial E}{\partial y(t+1)} \frac{\partial y(t+1)}{\partial u(t)} \frac{\partial u(t)}{\partial net_k} + wu(t) \frac{\partial O(t+1)}{\partial net_k} \qquad (4.7.19)$$

where $O(t+1) = \delta u(t)$ is the output of the neural network NN2.

We take an identity mapping as the activation function at the output layer. Then we have $O(t+1) = net_k$ and

$$\frac{\partial E}{\partial net_k} = -e(t+1) \frac{\partial y(t+1)}{\partial u(t)} + ru(t) \qquad (4.7.20)$$

since $\partial u(t)/\partial net_k = \partial O(t+1)/\partial net_k = 1$. Further, we define δ_k as $\delta_k = \partial E/\partial net_k$, then we have the following learning rule:

$$\Delta w_{kj}(t+1) = \eta \delta_k O_j + \Delta w_{kj}(t) \qquad (4.7.21)$$

$$\delta_k = - = e(t+1) \frac{\partial y(t+1)}{\partial u(t)} - ru(t). \qquad (4.7.22)$$

For the hidden layer, we have

$$\Delta w_{ji}(t+1) = \eta\delta_j O_i + \Delta w_{ji}(t) \tag{4.7.23}$$

$$\delta_j = w_{kj}\delta_k O_j(1-O_j). \tag{4.7.24}$$

Therefore, we can construct an optimal control strategy for this class of nonlinear systems.

4.7.6 Numerical Simulations

Consider the following plant given by

$$\begin{bmatrix} x_{m1}(t+1) \\ x_{m2}(t+1) \end{bmatrix} = \begin{bmatrix} 1 & 0 \\ 1 & 1 \end{bmatrix} \begin{bmatrix} x_1(t) \\ x_2(t) \end{bmatrix} + \begin{bmatrix} 1 \\ 0 \end{bmatrix} u(t)$$

$$x_1(t+1) = x_{m1}(t+1) + \eta(1 - \exp(x_{m2}(t+1)^2)$$

$$x_2(t+1) = x_{m2}(t+1) + \eta(1 - \exp(x_{m1}(t+1)^2)$$

$$y(t+1) = cx(t+1)$$

$$x(t+1) = \begin{bmatrix} x_1(t+1) \\ x_2(t+1) \end{bmatrix}$$

$$c = [0, 1] \tag{4.7.25}$$

where η denotes the degree of the plant nonlinearity. The cost function J is minimized as follows:

$$J = \frac{1}{2} \sum_{t=0}^{\infty} (y(t+1)^2 + 2u(t)^2). \tag{4.7.26}$$

If we assume the rough estimates for A and b as

$$A = \begin{bmatrix} 1.1 & -0.1 \\ 0.9 & 1.1 \end{bmatrix}, \quad b = \begin{bmatrix} 1.2 \\ 0.2 \end{bmatrix}, \tag{4.7.27}$$

then using LQ theory, we have

$$k = [0.977,\ 0.400] \tag{4.7.28}$$

for a finite-time settling observer, we obtain

$$f = \begin{bmatrix} 1.222 \\ 1.000 \end{bmatrix}. \tag{4.7.29}$$

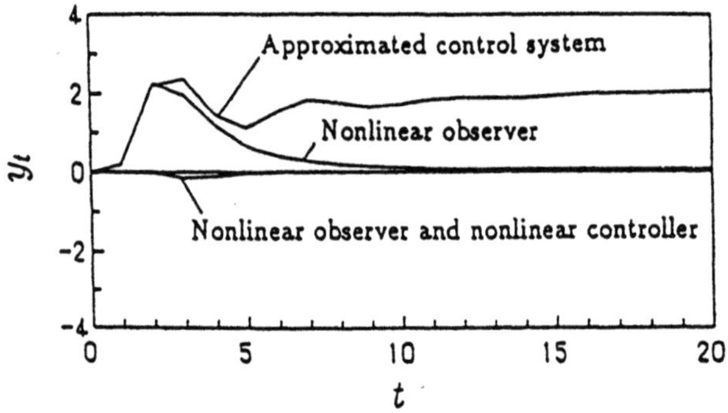

(a) Output of the plant.

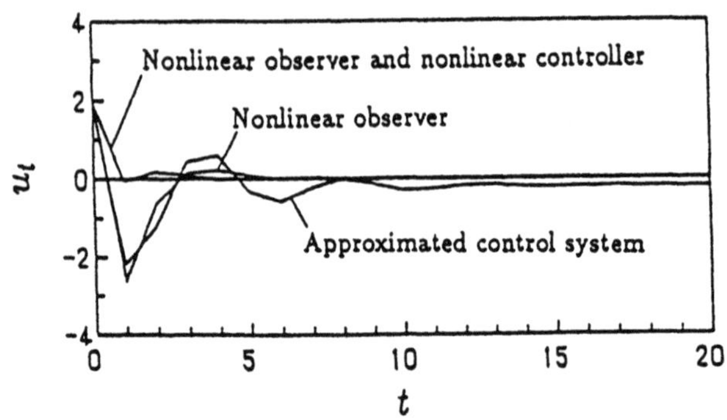

(b) Control input to the plant.

Fig. 4.7.3. Comparative results by various control algorithms.

Figure 4.7.3 (a) and (b) illustrate the control results by using the present approach. Figure 4.7.3 (a) shows the plant outputs whereas Fig.

4.7.3 (b) shows the plant inputs for the approximated linear controller case, nonlinear observer case, and nonlinear plus neuro-controller case. From these results, it can be observed that the neuro-control approach presented here is superior to the other traditional methods.

4.8 Self-Tuning Neuro-Control Scheme

The self-tuning neuro-control scheme is illustrated in Fig. 4.8.1 where a neural network is used to tune the parameters of a conventional controller similar to the adjustment made by a human operator.

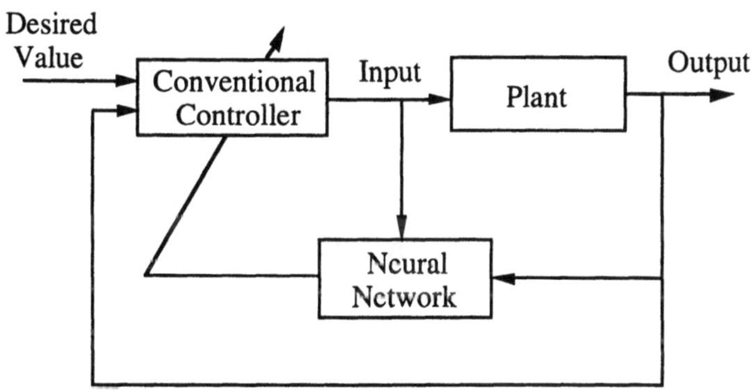

Fig. 4.8.1 A self-tuning type neuro-control scheme.

In this case, the human operator has accumulated some experience and knowledge on the control system, however, unlike a computer, it is rather impossible for the operator to store past data history of the system for any kind of operating conditions. A computer can store such information easily and retrieve it at once. Therefore, if we can include the operator's experience and knowledge into a neural network and train it based on past data history, then the trained neural network could be used as means to tune the controller parameters in an on-line way.

This approach has direct applications for many traditional control techniques which include adaptive control methods. Many adaptive control methods have a number of parameters or user-defined polynomials that are needed to be selected or tuned in prior. These are usually chosen by trial and error. By integrating a neural network into the control scheme, it can then be used to tune these parameters in an on-line way. Thus, this self-tuning neuro-control strategy has possible applications in many traditional control approaches. In the following section, we will discuss a self-tuning neuro-control scheme in which a neural network is used to tune the parameters of a PID controller referred to as the self-tuning PID neuro-control scheme.

4.9 Self-Tuning PID Neuro-Controller

PID controllers have a long history in control engineering and they have been proven to be robust, simple, and stable for many real world applications. Especially, in Japan, it has been reported that the percentage of PID controller that is currently being used in real control problems is about 84% [4]. This percentage shows that the PID controller is very much acceptable for many real applications due to its simplicity in structure and easily understandable principle.

The PID control configuration is illustrated in Fig. 4.9.1 where k_c is the proportional gain, T_i is an integral time, and T_d is the derivative time, which are adjusted to achieve the desired output. Note that for simplicity we have not considered the disturbances in Fig. 4.9.1.

The control input $u(\tau)$ of the PID controller is given by

$$u(\tau) = k_c\left(e(\tau) + \frac{1}{T_i}\int_0^{\tau} e(\tau')d\tau' + T_d \frac{de(\tau)}{d\tau} \right). \qquad (4.9.1)$$

Roughly speaking, the P action is related to the present error, the I action is based on the past history of error function, and the D action is related to the future behavior of the error. From estimation viewpoint, P, I, and D actions correspond to filtering, smoothing, and prediction problems, respectively.

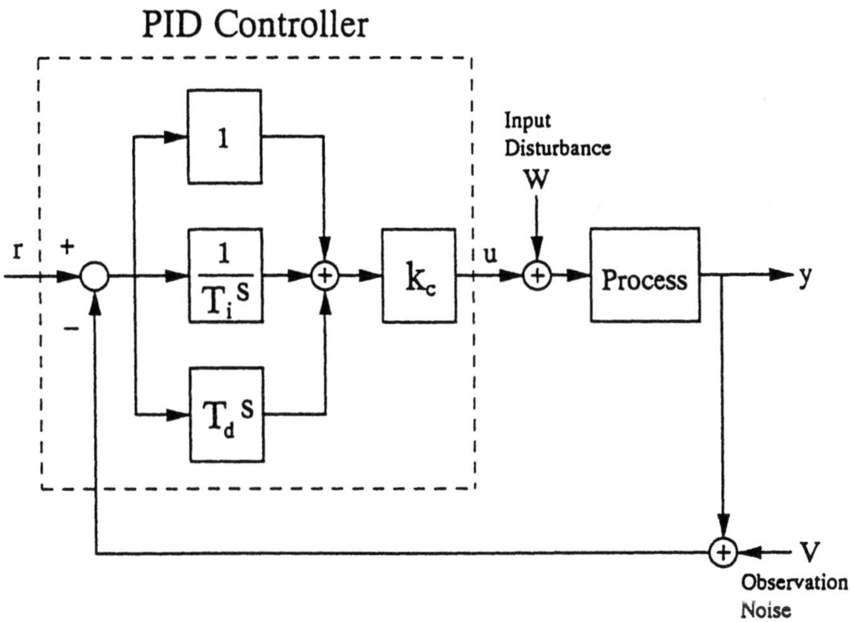

Fig. 4.9.1. General configuration of PID controller.

4.9.1 Derivation of the Self-Tuning PID Type Neuro-Controller

Figure 4.9.2 shows a general control structure for self-tuning PID learning where the neural network is used in place of human operators such that the error function is minimized by adjusting the PID gains [58]. The discrete-time version of PID controller is described by

$$u(t) = u(t-1) + K_P(e(t) - e(t-1)) + K_I e(t) + K_D(e(t) - 2e(t-1) + e(t-2)) \quad (4.9.2)$$

$$e(t) = r(t) - y(t)$$

where K_P, K_I, and K_D denote the PID gains and r(t) is a desired value of the plant output. In order to derive the self-tuning algorithm of the PID controller, we define the cost function E which should be minimized as

$$E = \frac{1}{2}e^2(t+1). \tag{4.9.3}$$

Using a three-layered neural network, we will realize the learning rule to find the suitable PID gains. Thus, the outputs at the output layer are K_P, K_I, and K_D which are denoted by O(1), O(2), and O(3), respectively. Based on the steepest descent (gradient) method, at the output layer

$$\Delta w_{kj}(t+1) = -\eta\frac{\partial E}{\partial w_{kj}} + \alpha\Delta w_{kj}(t) \tag{4.9.4}$$

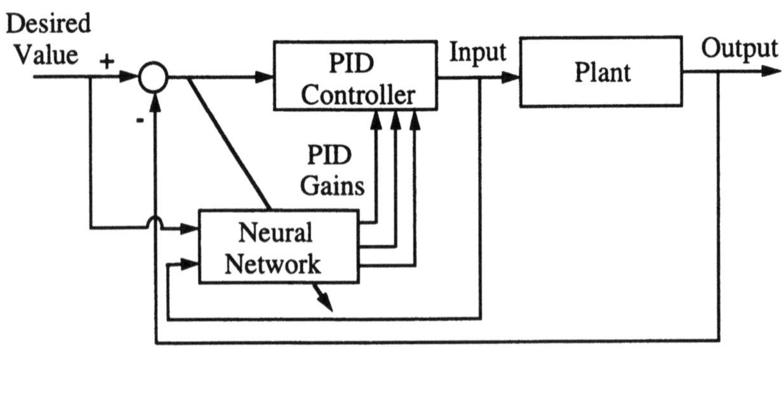

Fig. 4.9.2 Self-tuning PID learning configuration.

and at the hidden layer,

$$\Delta w_{ji}(t+1) = -\eta\frac{\partial E}{\partial w_{ji}} + \alpha\Delta w_{ji}(t). \tag{4.9.5}$$

Let us define δ_k by

$$\delta_k = -\frac{\partial E}{\partial net_k} \tag{4.9.6}$$

where

$$net_k = \sum_j w_{kj} O_j + \theta_k.$$

Let the output of neuron k at the output layer be denoted by O(k). Then we have

$$O(k) = f(net_k) \tag{4.9.7}$$

but the PID gains are not limited between 0 and 1. Thus, we can use some gain c after the output O(k) such that

$$O(k) = cf(net_k), \quad c > 0. \tag{4.9.8}$$

The parameter c can be also trained by increasing the hidden layer. For simplicity, we assume $c = 1$. Then using the chain rule, we have

$$\frac{\partial E}{\partial w_{kj}} = \frac{\partial E}{\partial net_k}\frac{\partial net_k}{\partial w_{kj}} = \frac{\partial E}{\partial net_k} O_j \tag{4.9.9}$$

and

$$\delta_k = -\frac{\partial E}{\partial net_k} = -\frac{\partial E}{\partial y(t+1)}\frac{\partial y(t+1)}{\partial u(t)}\frac{\partial u(t)}{\partial O(k)}\frac{\partial O(k)}{\partial net_k}. \tag{4.9.10}$$

But we have

$$\frac{\partial E}{\partial y(t+1)} = \frac{\partial E}{\partial e(t+1)}\frac{\partial e(t+1)}{\partial y(t+1)} = -(r(t+1) - y(t+1)) = -e(t+1)$$

$$\frac{\partial O(k)}{\partial net_k} = f'(net_k) = O(k)(1 - O(k))$$

$$\frac{\partial u(t)}{\partial O(k)} = \begin{cases} e(t) - e(t-1) & ; k = 1 \\ e(t) & ; k = 2 \\ e(t) - 2e(t-1) + e(t-2) & ; k = 3 \end{cases} \tag{4.9.11}$$

where the last relation has been obtained from (4.9.2) by noting that $O(1) = K_P$, $O(2) = K_I$, and $O(3) = K_D$.

Therefore, we have

$$\Delta w_{kj}(t+1) = \eta \delta_k O_j + \alpha \Delta w_{kj}(t), \tag{4.9.12}$$

where

$$\delta_k = e(t+1) \frac{\partial y(t+1)}{\partial u(t)} O(k)(1-O(k)) \frac{\partial u(t)}{\partial O(k)}. \tag{4.9.13}$$

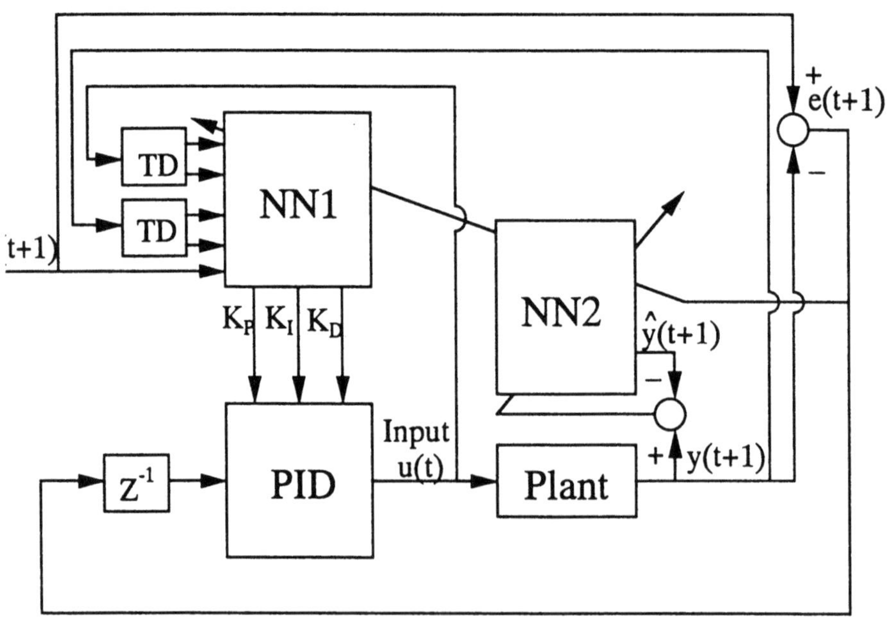

Fig. 4.9.3. Total configuration of the self-tuning PID type neuro-controller where NN1 is used to determine the PID gains and NN2 is used to determined the system's Jacobian.

For the hidden layer, we have

$$\frac{\partial E}{\partial net_j} = \sum_k \frac{\partial E}{\partial net_k}\frac{\partial net_k}{\partial O_j}\frac{\partial O_j}{\partial net_j} = -\sum_k \delta_k w_{kj} f'(net_j)$$

$$= -\sum_k \delta_k w_{kj} O_j(1-O_j).$$

Thus, we have

$$\Delta w_{ji}(t+1) = \eta\delta_j O_i + \alpha\Delta w_{ji}(t) \tag{4.9.14}$$

$$\delta_j = -\frac{\partial E}{\partial net_j} = \sum_k \delta_k w_{kj} O_j(1-O_j). \tag{4.9.15}$$

To calculate δ_k of (4.9.13), we need the system Jacobian $\partial y(t+1)/\partial u(t)$ and it has been estimated by using the emulator as before. The total configuration is illustrated in Fig. 4.9.3. Therefore, the self-tuning PID type neuro-control algorithm can be summarized by the following steps:

Step 1. Set the initial values of w_{kj}, w_{ji}, θ_k, θ_j, η and α. Start with $t = 0$ and go to Step 2.

Step 2. Calculate $e(t+1)$ and δ_k by

$$e(t+1) = r(t+1) - y(t+1)$$

$$\delta_k = e(t+1)\frac{\partial y(t+1)}{\partial u(t)} O(k)(1-O(k))\frac{\partial u(t)}{\partial O(k)}, \quad k = 1, 2, 3$$

where

$$\frac{\partial u(t)}{\partial O(k)} = \begin{cases} e(t) - e(t-1) & ; k = 1 \\ e(t) & ; k = 2 \\ e(t) - 2e(t-1) + e(t-2) & ; k = 3. \end{cases}$$

Step 3.

$$\Delta w_{kj}(n+1) = \eta\delta_k O_j + \alpha\Delta w_{kj}(n).$$

Step 4. Calculate δ_j by

$$\delta_j = \sum_k \delta_k w_{kj} O_j (1-O_j).$$

Step 5.

$$\Delta w_{ji}(t+1) = \eta \delta_j O_i + \alpha \Delta w_{ji}(t).$$

Step 6. $t \leftarrow t-1$ and go to Step 2.

Note that we use the emulator NN2 to calculate the system's Jacobian $\partial y(t+1)/\partial u(t)$ as shown in Fig. 4.9.3.

4.9.2 Simulation Examples

In this example we consider the following two models; one is a linear model (M1) and the other is a nonlinear model (M2):

$$\text{M1:} \quad y(t+1) = 0.988\, y(t) + 0.232u(t) \tag{4.9.16}$$

$$\text{M2:} \quad y(t+1) = 0.9\, y(t) - 0.001y(t-1)^2 + u(t) + \sin(u(t-1)). \tag{4.9.17}$$

The model M1 is obtained from the water bath temperature control system of Section 4.4.6 by using the least squares identification method. The model M2 is a modified nonlinear form of M1.

The performances of the two systems during the training phase are shown in Figs. 4.9.4 and 4.9.5 for M1 and M2, respectively, where the references are kept at 35 °C for M1 and 40 °C for M2. The training is done by repeating the desired trajectory cycles over thousands of times. The neural networks are used to minimize the performance error between the reference and the actual plant output. Convergence is achieved when the performance error falls below a pre-specified value. After training, it can be observed that the system perform well even for untrained references such that as shown in Figs. 4.9.6 and 4.9.7 for M1 and M2, respectively. It can also be observed that the self-tuning PID-type neuro-controller can track almost precisely even when the input signals are constrained i.e. simulating real control input conditions such that $u(t) \in [0, 5]$ for M1 and $u(t) \in [0,10]$

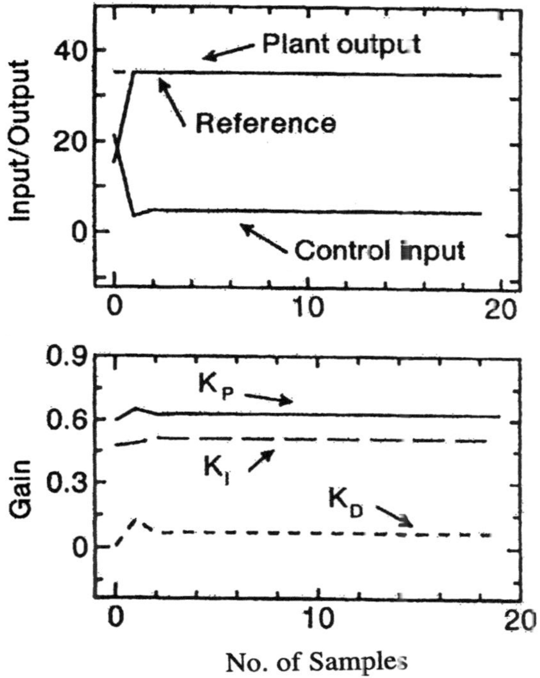

Fig. 4.9.4. Performance of system M1 during learning phase.

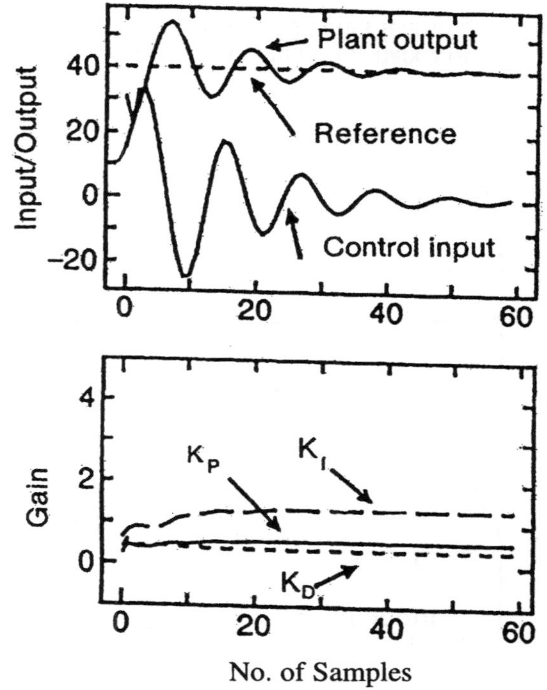

Fig. 4.9.5. Performance of system M2 during learning phase.

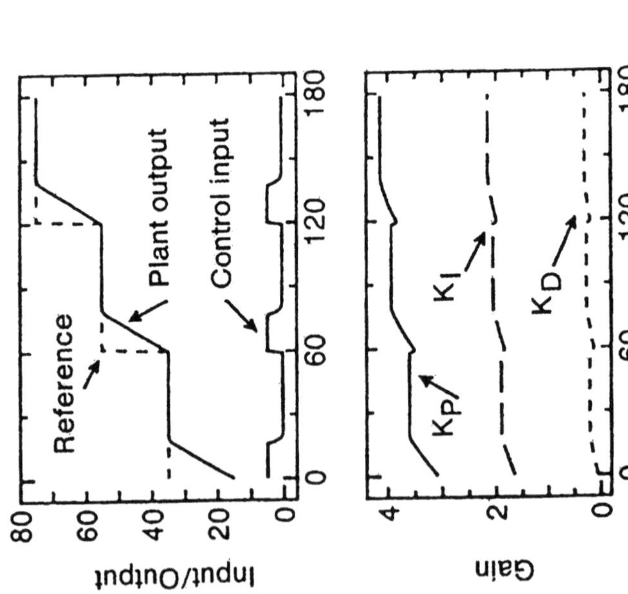

Fig. 4.9.6. Tracking performance of system M1 after learning.

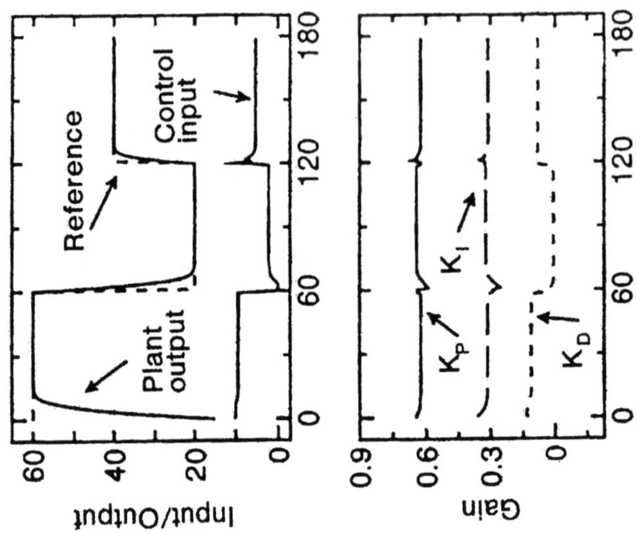

Fig. 4.9.7. Tracking performance of system M2 after learning.

for M2. The real application example on the water bath temperature control using another neuro-control method will be described in the next chapter.

4.10 Emulator and Controller Neuro-Control Scheme (Backpropagation-Through-Time Scheme)

The neuro-control scheme of Fig. 4.2.5 is basically an integration and modification of the direct inverse model neuro-control scheme (Fig. 4.2.2) and the specialized learning scheme (Fig. 4.2.3). In this approach the Jacobian or partial derivatives of the plant $\partial y / \partial u$ can be approximated by the neural network emulator instead of approximating it by using the signs of the Jacobian. The scheme initially involves off-line training of neural network models to learn the inverse and forward neural network models of the plant. The neural network model that is trained to learn the plant inverse is used as a controller whereas the neural network model that is trained to learn the plant forward dynamics is used as means to backpropagate the performance error of the system to obtain the equivalent error at the output of the neuro-controller. The derivation of the error signals is described in the sections that follow.

4.10.1 Off-Line Training of the Neuro-Controller and Emulator

Similar to the methodology of the direct inverse model neuro-control as described earlier, the neuro-controller models are trained to learn the inverse dynamic model of the plant by using the plant output data together with some corresponding delayed output values as the input patterns and the corresponding plant input data as the target patterns. The neuro-emulator models are trained to learn the forward dynamics of the plant by using the input data together with some delayed output data as the input patterns and the corresponding output data as the target patterns. In training a neural network to learn an inverse or a forward dynamic model of a plant off-line, the back-error-propagation signal between the output and the hidden layers, is expressed as

$$\delta_k = \tau_k - O_k \tag{4.10.1}$$

where τ_k is the target pattern and O_k is the actual output of the neural network, and between the hidden and input layers, it is expressed as

$$\delta_j = f'_k(net_j)\sum_k \delta_k w_{kj}. \tag{4.10.2}$$

Here, f' is the derivative of the functions of f where f(x)= 1/(1+exp(-x)). The weights between the input and hidden layers are updated as

$$\Delta w_{ji}(t+1) = \eta\delta_j O_i + \alpha\Delta w_{ji}(t) + \beta\Delta w_{ji}(t-1) \tag{4.10.3}$$

and between the hidden and output layers as

$$\Delta w_{kj}(t+1) = \eta\delta_k O_j + \alpha\Delta w_{kj}(t) + \beta\Delta w_{kj}(t-1) \tag{4.10.4}$$

where O_i and O_j are the outputs of the input and hidden layers, respectively, η is the learning rate, and α and β are the momentum and acceleration coefficients, respectively. The constants η, α, and β are all chosen empirically. (4.10.3) and (4.10.4) are modified forms of the generalized delta rule [56].

4.10.2 On-line Learning

To illustrate the derivation of the error signals for the neuro-controller Fig. 4.10.1 is referred. For this on-line scheme, adaptation of the weights of the controller neural network between the hidden and output layers can be derived as follows

$$\Delta w^C_{kj} \propto \frac{\partial E^E}{\partial w^C_{kj}} = \eta\frac{\partial E^E}{\partial w^C_{kj}} \tag{4.10.5}$$

where

$$E = \frac{1}{2}(r-y)^2$$

where r and y are the desired and actual plant outputs, respectively. The superscript C denotes the variables belonging to the controller neural network and superscript E denotes the variables belonging to the emulator neural network.

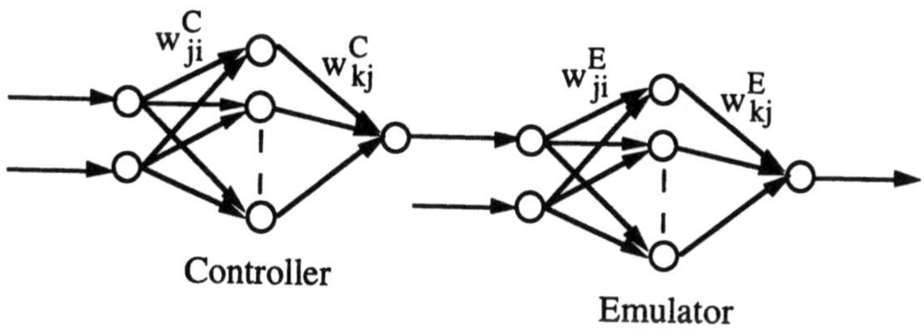

Fig. 4.10.1. Figure showing the connection weights of the emulator and controller neural networks.

Using chain rule, (4.10.5) can be expanded as follows

$$\Delta w_{kj}^C = \eta \frac{\partial E^E}{\partial net_k^C} \frac{\partial net_k^C}{\partial w_{kj}^C} = \eta \delta_k^C O_j^C \tag{4.10.6}$$

where

$$\frac{\partial E^E}{\partial net_k^C} = \delta_k^C \quad \text{and} \quad \frac{\partial net_k^C}{\partial w_{kj}^C} = O_j^C .$$

From this point it would be more relevant to show how the error signals between the layers of the neural networks are derived. δ_k^C is the error signal between the hidden and output layers of the controller neural network. We use linear functions at the input and output neurons of the neural networks, therefore we can equate as follows

$$\delta_k^C = \frac{\partial E^E}{\partial O_k^C} = \frac{\partial E^E}{\partial net_i^E} = \frac{\partial E^E}{\partial O_i^P} \tag{4.10.7}$$

where net_i^E and O_i^E are the input and output of the emulator input layer neurons. Further using chain rule

$$\delta_k^C = \frac{\partial E^E}{\partial net_j^E} \frac{\partial net_j^E}{\partial O_i^E}. \tag{4.10.8}$$

Here we have

$$\frac{\partial E^E}{\partial net_j^E} = \delta_j^E \text{ and } \frac{\partial net_j^E}{\partial O_i^E} = w_{ji}^E$$

hence,

$$\delta_k^C = \delta_j^E w_{ji}^E. \tag{4.10.9}$$

The error signal between the input and hidden layer of the controller neural network can be derived as follows

$$\delta_j^C = \frac{\partial E^E}{\partial net_j^C} = \frac{\partial E^E}{\partial O_j^C} \frac{\partial O_j^C}{\partial net_j^C}. \tag{4.10.10}$$

By chain rule

$$\frac{\partial E^E}{\partial O_j^C} = \frac{\partial E^E}{\partial net_k^C} \frac{\partial net_k^C}{\partial O_j^C}$$

and

$$\frac{\partial O_j^C}{\partial net_j^C} = f'\left(net_j^C\right)$$

where

$$f'\left(net_j^C\right) = O_j^C\left(1 - O_j^C\right),$$

hence,

$$\delta_j^C = \delta_k^C w_{kj}^C O_j^C \left(1 - O_j^C\right) \tag{4.10.11}$$

In this scheme, the weights of the emulator neural network are not kept constant but are further improved on-line. This is done by backpropagation of the following error equation through the emulator neural network at every sample where y and $\hat{y}$ are the outputs of the actual plant and emulator, respectively

$$\varepsilon = \frac{1}{2}(y - \hat{y})^2. \tag{4.10.12}$$

This emulator and controller neuro-control scheme is applied to a water bath process and a MIMO furnace which is described in Chapter 5.

4.11 Conclusions

In this chapter an overview of neuro-control approaches have been discussed. From the authors' viewpoint, five neuro-control schemes have been classified. Four of the schemes which are based on the backpropagation algorithm namely; series control scheme, parallel control scheme, self-tuning control scheme, and emulator and controller scheme; have been discussed and derived. Extensions of the series and parallel neuro-control schemes have also been developed. Simulation examples have been given for several of the neuro-control schemes.

4.12 References

[1] Åström, K.J. and B. Witternmark, *"Adaptive Control,"* Addison-Wesley, New York, 1989.

[2] Box, G.E. and G.H. Jenkins, *"Time Series Analysis: Forecasting and Control"*, Holden-Day, San Francisco, 1976.

[3] Kosko, B., *"Neural Networks for Signal Processing"*, Prentice-Hall, Englewood Cliffs, New Jersey, 1992.

[4] Nozaka, Y., "Trend of new control theory application in industrial process control (A survey)", *Proc. of 12th IFAC World Congress*, Sydney, Vol. VI, pp. 51-56, 1993.

[5] Åström, K.J., "Towards intelligent control", *IEEE Control Systems Magazine,* Vol. 9, pp.60-69, 1989.

[6] White, D.A. and D.A. Sofge, *"Handbook of Intelligent Control"*, Van Nostrand Reinhold, New York, 1992.

[7] Wiener, N., *"Cybernetics or Control and Communication in the Animal and the Machine"*, MIT Press, Cambridge, Massachusetts, 1948.

[8] Cybenko, G., "Approximation by superpositions of a sigmoidal function", *Mathematics Control, Signal & System*, Vol. 2, pp.303-314, 1989.

[9] Funahashi, K.I., "On the approximate realization of continuous mappings by neural networks", *Neural Networks*, Vol. 2, pp.183-192, 1989.

[10] Hornik, K., M. Stinchcombe, and H. White, "Multilayer feedforward networks are universal approximators", *Neural Networks*, Vol. 2, pp. 359-366, 1989.

[11] Saerens, M. and A. Soquet, "A neural controller based on backpropagation algorithm", *Proc. of First IEE Int. Conf. on Artificial Neural Networks*, London, pp. 211-215, 1989.

[12] Ogata, K., *"Discre-time Control Systems"*, Prentice-Hall, Englewood Cliffs, New Jersey, 1987.

[13] Iiguni, Y., H. Sakai, and H. Tokumaru, "A non-linear regulator design in the presence of system uncertainties using multilayered neural networks,"*IEEE Trans. on Neural Networks,* Vol. 2, pp. 410-417, 1991.

[14] Narendra, K.S. and K. Parthasarathy, "Identification and control of dynamical systems using neural networks", *IEEE Trans. on Neural Networks*, Vol. 1, pp. 4-27, 1990.

[15] Neuralogix NLX420 Datasheet, American Neuralogix Inc., 1992.

[16] Levin, E., R. Gewirtzman, and G.F. Inbar, "Neural network architecture for adaptive system modelling and control", *Proc. of Int.*

Joint Conf. on Neural Networks, Washington D.C., Vol. II, pp. 311-316, 1989.

[17] Kosko, B., *"Neural Networks and Fuzzy Systems: A Dynamical Systems Approach to Machine Intelligence"*, Prentice-Hall, Englewood Cliffs, New Jersey, 1991.

[18] Miller, W.T., R.S. Sutton, and P.J. Werbos, *"Neural Networks for Control"*, MIT Press, Cambridge, MA, 1990.

[19] Barto, A.G., "Connectionist learning for control", in *Neural Networks for Control*, MIT Press, Cambridge, Massachusetts, pp. 5-58, 1990.

[20] Werbos, P.J., "Overview of designs and capabilities", in *Neural Networks for Control*, MIT Press, Cambridge, MA, pp. 59-65, 1990.

[21] Hunt, K.J., D. Sbarbaro, R. Zbikowski, and P.J. Gawthrop, "Neural networks for control systems - a survey", *Automatica*, Vol. 28, pp. 1083-1112, 1992.

[22] Widrow, B. and F.W. Smith, "Pattern-recognizing control systems", *Proc. of Computer and Information Sciences*, Washington D.C., Spartan, Washington, 1964.

[23] Albus, J.S., "A new approach in manipulator control: the cerebellar model articulation controller (CMAC)", *Journal of Dynamic Systems, Measurement and Control*, pp. 220-227, 1975.

[24] Albus, J.S., "Data storage in the cerebellar model articulation controller", *Journal of Dynamic Systems, Measurement and Control*, pp. 228-233, 1975.

[25] Miller III, W.T., F.H. Glanz, and L.G. Kraft, "CMAC: An associative neural network alternative to backpropagation", *Proc. of IEEE*, Vol. 78, pp. 1561-1567, 1990.

[26] Barto, A.G., R.S. Sutton, and C.W. Anderson, "Neuronlike adaptive elements that can solve difficult learning control problems", *IEEE Trans. on Systems, Man, and Cybernetics*, Vol. SMC-13, pp.834-846, 1983.

[27] Gaines, B.R. and J.H. Andreae, "A learning machine in the context of the general control problem", *Proc. of 3rd IFAC Congress*, IME, London, pp. 14B.1-14B.8, 1966.

[28] Witten, I.H., "An adaptive optimal controller for discrete-time Markov environments", *Information and Control*, Vol. 34, pp. 286-95, 1977.

[29] Werbos, P.J., "Advanced forecasting methods for global crisis warning and models of intelligence", *General System Yearbook*, Vol. 22, pp.25-38, 1977.

[30] Holland, J.H., "Escaping brittleness: The possiblity of general-purpose learning algorithm algorithms applied to rule-based systems", R.S. Michalski, J.G. Carbonell, and T.M. Mitchell, Eds., *Machine Learning: An Artificial Intelligence Approach, Vol. II*, Morgan Kaufmann, San Mateo, CA, 1986.

[31] Anderson, C.W., "Learning and problem solving with multilayer connectionist systems", *Ph.D. Thesis,* University of Mass., Amherst, 1986.

[32] Anderson, C.W., "Strategy learning with multilayer connectionist representations", *Tech. Rep. TR87-509.3*, GTE Laboratories Inc., Waltham, MA, 1987.

[33] Widrow, B. and S. D. Stearns,"*Adaptive Signal Processing*", Prentice-Hall, Englewood Cliffs, New Jersey, 1985.

[34] Kuperstein, M., "Neural model of adaptive hand-eye co-ordination for single postures", *Science*, Vol. 239, pp. 1308-1311, 1988.

[35] Kuperstein, M., and J. Rubinstein, "Implementation of an adaptive neural controller for sensory-motor co-ordination", *IEEE Control Systems Magazine,* Vol. 9, pp. 25-30, 1989.

[36] Srinivasan, V., A.G. Barto, and B.E. Ydstie,"Pattern recognition and feedback via parallel distributed processing", *Annual Meeting of the American Institute of Chemical Engineers*, Washington, D.C., 1988.

[37] Sanner, R.M. and D.L. Akin, "Neuromorphic pitch attitude regulation of an underwater telerobot", *IEEE Control Systems Magazine,* Vol. 10, pp.62-68, 1990.

[38] Psaltis, D., A. Sideris, and A. Yamamura, "A Multilayered neural network controller", *IEEE Control Systems Magazine,* Vol. 8, pp.17-21, 1988.

[39] Saerens, M. and A. Soquet, "A neural controller based on backpropagation algorithm", *Proc. of First IEE Int. Conf. on Artificial Neural Networks*, London, pp. 211-215, 1989.

[40] Kawato, M., K. Furukawa, and R. Suzuki, "A hierarchical neural network model for control and learning of voluntary movement", *Biological Cybernetics*, Vol. 57, pp. 169-185, 1987.

[41] Kawato, M., Y. Uno, M. Isobe, and R. Suzuki, "Hierarchical neural network model for voluntary movement with application to robotics", *IEEE Control Systems Magazine,* Vol. 8, pp.8-16, 1988.

[42] Jordan, M.I., "Generic constraints on underspecified target trajectories", *Proc. of Int. Joint Conf. on Neural Networks (IJCNN)' 89*, Washington, Vol. I, pp. 217-225, 1989.

[43] Jordan, M.I. and D.E. Rumelhart, "Forward models: Supervised learning with a distal teacher", *Cognitive Science*, Vol. 16, pp.313-355, 1990.

[44] Narendra, K.S. and K. Parthasarathy, "Identification and control of dynamical systems using neural networks", *IEEE Trans. on Neural Networks*, Vol. 1, pp. 4-27, 1990.

[45] Nguyen, D.H. and B. Widrow, "Neural networks for self-learning control systems", *IEEE Control Systems Magazine,* Vol. 10, pp. 18-23, 1990.

[46] Chen, S., S.A. Billings, and P.M. Grant, "Nonlinear system identification using neural networks", *Int. Journal of Control*, Vol. 51, pp. 1215-1228, 1990.

[47] Chen, S., S.A. Billings, C.F. Cowan, and P.M. Grant, "Practical identification of NARMAX models using radial basis function", *Int. Journal of Control*, Vol. 52, pp. 1327-1350, 1990.

[48] Bhat, N.V., Jr., P.A. Minderman, T. McAvoy, and N.S., Wang, "Modeling chemical process systems via neural computation", *IEEE Control Systems Magazine,* Vol. 10, pp. 24-30, 1990.

[49] Chu, S.R., R. Shoureshi, and M. Tenorio, "Neural networks for system identification", *IEEE Control Systems Magazine,* Vol. 10, pp. 31-35, 1990.

[50] Billings, S.A., H.B. Jamaludin, and S. Chen, "Properties of neural networks with applications to modeling of nonlinear dynamical systems", *Int. Journal of Control*, Vol. 55, pp. 193-224, 1992.

[51] Sjoberg, J. and L. Ljung, "Overtraining, regularization, and searching for minimum in neural networks", *Revised Version Report Lith-ISY-I-1297*, Dept. of Elec. Eng., Linkoping University, Sweden, 1992.

[52] Tanomaru, J. and S. Omatu, "Process control by on-line trained neural controllers", *IEEE Trans. on Industrial Electronics,* Vol. 39, pp. 511-521, 1992.

[53] Hopfield, J.J., "Neural networks and physical systems with emergent computational abilities", *Proc. of the National Academy of Sciences*, Vol. 79, pp. 2554-2558, 1982.

[54] Lippmann, R., "An introduction to computing with neural nets" *IEEE ASSP Magazine*, Vol. 4, pp.4-22, 1987.

[55] Gomi, H. and M. Kawato, "Neural network control for a closed loop system using feedback error learning", *Neural Networks*, Vol. 6, pp. 933-946, 1993.

[56] Nagata, S., M. Sekiguchi, and K. Asakawa, "Mobile robot control by a structured hierarchical neural network", *IEEE Control Systems Magazine,* Vol. 10, pp. 69-76, 1990.

[57] Kinjo, H., S. Omatu, T. Yamamoto, and S. Tamaki, "Suboptimal control for a non-linear system using neural networks", *Proc. of 1st Asian Control Conference*, pp. 551-554, Tokyo, 1994.

[58] Saiful A. and S. Omatu, "Neuromorphic self-tuning PID controller", *Proc. of 1993 IEEE ICNN*, San Francisco, pp. 552-557, 1993.

CHAPTER 5
NEURO-CONTROL APPLICATIONS

5.1 Introduction

In this chapter we discuss several neuro-control techniques with applications to real physical processes; a water bath temperature control system, an inverted pendulum, an electric vehicle generator control system, and a multi-input multi-output furnace. For the water bath and furnace temperature control systems, the emulator and controller neuro-control scheme is applied. However, as these real processes are slow in nature, off-line learning methods are used to train the neural networks at first and then on-line learning is applied using the architecture of Fig. 4.2.5 for fine-tuning their performances. In these applications comparison is made with several traditional control methods under varying complexities in the processes. As neuro-control is relatively new it is important to see how well it compares to the more established traditional control approaches.

For the inverted pendulum, fuzzy logic is first used as an initial start-up control using only a minimum set of rules. This kind of control can be observed as a coarse control action as fuzzy logic is rather easily applied. However, choosing the correct set of rules and scale factors is not an easy task in order to fine-tune the fuzzy controller for optimum performance. In this case linear quadratic optimal control strategy is used to stabilize the pendulum in an upright manner. Finally, neural networks are used for fine improvement of the two controllers to overcome nonlinearity and unknown dynamics in the system.

Another application is in the development of the self-tuning PID type neuro-controller on the generator system of an electric vehicle. A neural network is used to tune the gains of a PID controller in an on-line way. Experiments show that the final error could be reduced using the neural network approach of tuning the PID controller gains.

5.2 Application of Neuro-Control to a Water Bath Process and Comparison with Alternative Control Schemes

5.2.1 Introduction

Classical feedback control theory has been the basis for the development of simple automatic control systems. Its easily comprehensible principle and relatively simple implementation have been the main reason for its wide applications in industry. However, such fixed-gain feedback controllers are insufficient to compensate for parameter variations in the plant as well as to adapt to changes in the environment. The need to overcome such problems and to have a controller well-tuned not just for one operating point but for a whole range of operating points has motivated the idea of an adaptive controller.

In the last three decades, numerous adaptive control techniques [1], [2], have been proposed to replace the conventional classical methods. These developments in adaptive control theory were largely due to significant developments in modern control and stability theories. The ability to adapt to variations in the plant dynamics and environment automatically has made such adaptive controllers increasingly important for various applications. However, before they can be implemented, mathematical modeling of the plants has to be done, which sometimes is difficult and laborious. In addition, inaccuracy in the modeling of the plants could lead to degraded performances of the controllers. They also contain many control parameters which made it difficult for field engineers and process operators to relate the parameter settings and control system actions. Moreover, their applications are not suited to nonlinear and complex control problems.

The advent of artificial neural networks and fuzzy logic theory has inspired new resources for the possible implementation of better and more efficient control. Neural networks are trainable dynamical systems that estimate input-output functions. Fuzzy systems transform sets of structured information regarding the systems into the appropriate control actions. Both of these systems encode sampled information in a parallel-distributed framework and can be feasibly implemented in hardware [3]-[5]. Neural and fuzzy systems offer a key advantage over traditional adaptive and optimal control systems. They do not require *a priori* mathematical models of the plants for implementation. Using a few common sense associations of how the control variables behave, fuzzy control systems can be developed along linguistic lines which have close association with the field of artificial intelligence. Neural networks on the other hand have to be trained in prior before they can be used for control purposes.

There have been some studies recently where neural network based control systems are compared to fuzzy and several adaptive control systems [6], [7]. However, these studies are based entirely on simulations. In this research, we compare the performance of a multilayered neuro-controller (NNC) to a fuzzy logic controller (FLC), a generalized predictive controller (GPC), and a conventional PI-controller (CFC) in regulating the temperature of a water bath control system.

Why do we need to compare different control algorithms? Different control algorithms use different approaches, therefore, by applying each algorithm to solve the same control problem we would be able to understand their differences and similarities. We would be able to decipher the advantages and disadvantages of each control approach by observing how well they performed under identical conditions. Moreover, the neuro-control approach is still in its infancy and its application in real-time control is still far behind the more established traditional and modern control methods. Thus, it would be interesting to see how well it performs and compares to the more established control approaches.

How do we compare these different control algorithms? It would be rather impossible to compare linear controllers such as the GPC and the CFC with nonlinear controllers such as the NNC and the FLC on complex ill-defined nonlinear control problems. Hence, this study centers on a simple water bath system. Direct comparison among the four algorithms provides a quick and simple analysis of each methodology and their respective performances. In order to make identical comparisons among the algorithms, experiments are carried out where each system is subjected to the same artificial disturbances and variations in the plant.

Four groups of experiments are carried out for each system. The four systems are compared with respect to set-point regulation, the effect of unknown load disturbances, large parameter variation, and variable time-delay in the system. The implementation of each algorithm requires local adjustments and heuristic corrections on a particular plant. Although it is naive to make strong conclusions on the relative merits and demerits of each system, it is hoped that the results of the experimental studies will provide some insights on the future development of an efficient and intelligent control approach. Research efforts should be directed toward extracting the best characteristics from each of these different classes of systems.

In the neuro-control approach, multilayered neural networks are first trained off-line using the backpropagation algorithm to learn the inverse and forward dynamic models of the plant. The neural networks are then configured in the architecture as shown in Fig. 4.2.5 where they can be continuously trained in an on-line way to further improve their performance. This architecture has been applied independently by a few researchers [8]-[10], where the NNC is trained from initial random weights. However, this approach is not practical for this slow varying process in real-time and hence, it is necessary to train both the inverse and forward neural network models initially off-line.

The FLC system is implemented using simple common sense associations regarding the temperature control of the process. Although in recent years adaptive fuzzy control systems have been proposed where the rules can be automatically composed and decomposed at the expense of added complexity in developing the systems (some examples are [7], [11]-[13]), it is our intention to use the conventional fuzzy approach based on [27]. The generalized predictive control system is implemented using the long range predictive control algorithm developed by Clarke, Mohtadi, and Tuffs [14], [15] and has been discussed in length in Section 3.6. As PI or

PID control is still by far the most widely used method in industrial applications, it is also worthwhile to compare this conventional scheme to the other three control schemes under identical conditions on the same single-input-single-output system.

This section has been organized as follows. Firstly, a brief description of the water bath temperature control system is presented. In the following sections, the implementation of neuro-control, fuzzy logic control, and generalized predictive control on the water bath system are briefly explained. The results of the experimental studies are compared and discussed in the sections that follow.

5.2.2 Description of the Water Bath Temperature Control System

The temperature control system consists of a Yamato Science Inc. laboratory water bath temperature control process (BT-15 model). A schematic diagram of the experimental setup is shown in Fig. 5.2.1. The system can be divided into five main components; (i) the water bath, (ii) the sensor module, (iii) the programmable input-output (PIO) interface board, (iv) the microcomputer, and (v) the actuator. A brief description of the five components follows. The capacity of the water bath is 8 liters with dimension 250x290x100 (mm^3). The water bath is heated by a 600W base heater which is connected to a thyristor (SI6G12S-12) circuit. To ensure even temperature distribution, a stirrer is used which can rotate at 120 rpm. The sensor module has been developed using diodes (1S1588) and high gain amplifiers (A741). It consists of a two step amplification circuit and can transform the measured temperature over the range of 0°C to 100°C into the corresponding voltage range of 0V to 10V with a resolution of 0.24 °C.

The PIO interface circuit board consists of an analog-to-digital (A/D) converter, a digital-to-analog (D/A) converter, and a programmable peripheral interface device (PD8255A). An external clock is designed to operate the A/D and D/A converter. The clock circuit is designed using a crystal oscillator and JK flip flops. The microcomputer used in this experiment is the NEC PC 9801F having an Intel 8086 16-bit CPU with a 10 MHz clock speed. A simple control routine is written using Microsoft-C to provide the control input to the actuator through the D/A and also to measure the output temperature.

A thyristor (S16G12S-12) is used as an actuator for the heater and is switched on and off according to the following constraints;

if $u(tT) \leq 0.0$, then $V_i = 0.0V$, heater off

if $u(tT) \geq 5.0$, then $V_i = 5.0V$, heater on

and if $0.0 < u(tT) < 5.0$, then $V_i = u(tT)V$, heater on.

where u(tT) is the output of the controller, T denotes the sampling interval, t is the sampling number (t=0,1,2,3,...), and V_i is the input voltage to the actuator.

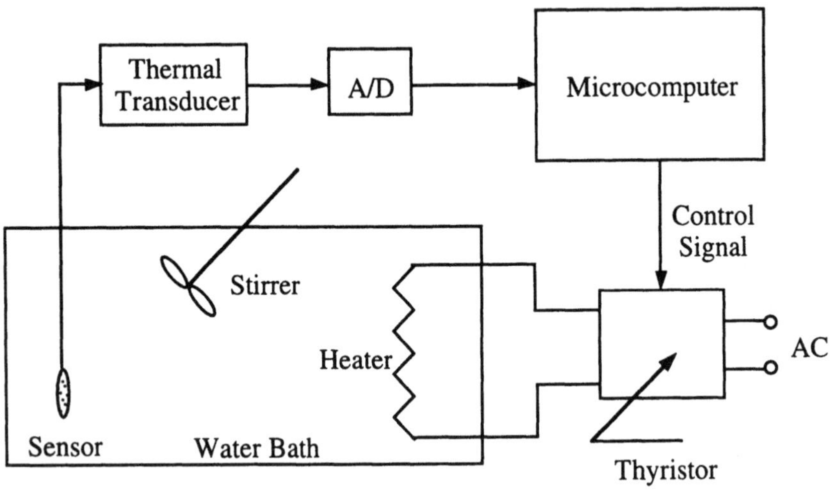

Fig. 5.2.1. Schematic diagram of the water bath temperature control system.

5.2.3 Neuro-Control Scheme

The ability of artificial neural networks to learn and generalize has found wide interest in a wide variety of applications. One of the simplest approaches for the implementation of neuro-control is the direct inverse control approach pioneered by Widrow and Stearns [16], and Psaltis *et. al.* [17]. In the Psaltis *et. al.*'s scheme, referred to as general learning as discussed in Section 4.2.1, a multilayered neural network is first trained off-line using backpropagation algorithm [18] to learn the inverse dynamic model of a plant and once trained it is configured as a direct controller to the plant. In a simple study by the authors [19], it was found that the number of delayed plant signals as input vector elements of the neural network model used in the training stage did have an effect on the neuro-controller in learning the true inverse plant model which subsequently affected the performance of the neuro-controller on-line. Theoretically, this can be described simply as follows.

Consider a single-input single-output linear dynamic plant, without noise, which has the following discrete mathematical model

$$y(t) + a_1 y(t-1) + \ldots + a_q y(t-q) = z^{-d}\left[b_0 u(t) + b_1 u(t-1) + \ldots + b_p u(t-p)\right] \tag{5.2.1}$$

where y and u are the plant output and control input, respectively. Here, a_i i=1,2,...q are the denominator coefficients, b_i i=1,2,...p are the numerator coefficients, d is the dead-time or time-delay, and t is the sample number. The neural network model for the inverse dynamic model of the plant is a nonlinear function g, which consists of the delayed plant outputs and inputs according to the estimated order and time-delay of the plant:

$$\begin{aligned} I(t) &= g\{u(t-d)\} \\ &= g\{y(t), y(t-1), \ldots, y(t-q), u(t-d-1), \ldots, u(t-d-p)\} \end{aligned} \tag{5.2.2}$$

where I(t) is the output of the neural network which is the actual control input to the plant. In implementing the neuro-control system on-line the actual plant output signal y(t) is replaced by the desired output r(t) as one of the elements of the neural network input vector.

Based on this simple formulation, we developed the inverse and forward neural network models of the plant from a knowledge of its reduced order mathematical model [20]. Although it has been shown that neural networks can be trained to learn plant models without any prior mathematical modeling [8]-[10], it would be sophistry to ignore any information regarding the plant in designing controllers when such information is available. Moreover, based on such information, unnecessary trial and error procedures can be eliminated [19].

The inverse neural network model has been trained using a selection of training patterns taken from the open-loop plant response. In a number of applications [8]-[10], [21], the training patterns are selected by probing the plant input using random signals. In real-time, this technique is rather impractical for this slowly varying temperature control process. As it is open-loop stable, a ramp signal was instead directly injected to the actuator of the plant within the limits of the actuator input constraints. A set of corresponding input-output training patterns was then selected from the open-loop ramp response of the process as shown in Fig.5.2.2.

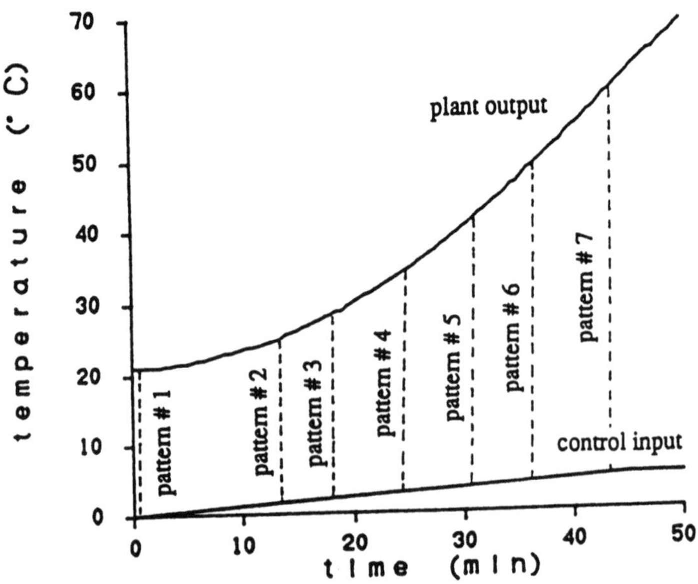

Fig. 5.2.2. Open-loop input-output characteristic of the water bath process by using a ramp signal injected directly to the actuator. A set of seven input-output training patterns are selected from the input-output characteristic.

It was found that the neural network model having eight hidden neurons could learn the true inverse plant model using two input vector elements which consist of the present plant output, y(t) and one delayed output, y(t-1) based on the plant's reduced-order mathematical model. The forward neural network model was similarly developed, however, in this case, the input training patterns consist of a selection of the plant input and delayed output signals and the output patterns consist of the corresponding actual plant output. Both the neural network models are configured as in the architecture of Fig. 4.2.5 where on-line learning can be done to further improve their performances.

For the off-line general learning scheme, the backpropagation signal between the output and the hidden layers, is expressed as

$$\delta_k = \tau_k - O_k \tag{5.2.3}$$

where τ_k is the target pattern and O_k is the actual output of the neural network, and between the hidden and input layers, it is expressed as

$$\delta_j = f'(net_j)\sum_k \delta_k w_{kj} . \tag{5.2.4}$$

Here, $f'(net_j)$ is the derivative of the function $f(net_j)$ where

$$f(net_j) = 1/(1+\exp(-net_j)) .$$

The connection weights between the input and hidden layers are updated as

$$\Delta w_{ji}(t+1) = \eta\delta_j O_i + \alpha\Delta w_{ji}(t) + \beta\Delta w_{ji}(t-1) \tag{5.2.5}$$

and between the hidden and output layers as

$$\Delta w_{kj}(t+1) = \eta\delta_k O_j + \alpha\Delta w_{kj}(t) + \beta\Delta w_{kj}(t-1) \tag{5.2.6}$$

where η is the learning rate, and α and β are the momentum and acceleration coefficients, respectively. Equations (5.2.5) and (5.2.6) are modified forms of the generalized delta rule [22].

For the on-line scheme, the connection weights of the neuro-controller are adapted using the performance error as follows:

$$E^E = \frac{1}{2}[r - y]^2 \tag{5.2.7}$$

where r and y are the desired and the actual plant outputs, respectively. This error is backpropagated through the emulator. Thus, the error signal at the output of the controller can be derived as follows:

$$\delta_k^C = -\frac{\partial E^E}{\partial net_k^C} = -\frac{\partial E^E}{\partial O_k^C}\frac{\partial O_k^C}{net_k^C} \tag{5.2.8}$$

$$\delta_k^C = \sum_j \delta_j^E w_{ji}^E \tag{5.2.9}$$

where O_k^C and net_k^C are the output and input of the controller's output layer neurons, respectively. The superscript E denotes the emulator and C denotes the controller. The error signal between the hidden and input layers of the controller is

$$\delta_j^C = \left(\sum_j \delta_k^C w_{kj}^C\right) O_j^C \left(1 - O_j^C\right) \tag{5.2.10}$$

where O_j^C is the output of the controller's hidden layer neurons, respectively.

In this scheme, the connection weights of the neuro-emulator are not kept constant but are further improved on-line. This is done by backpropagation of the following error E through the emulator at every sample where y and $\hat{y}$ are the outputs of the actual plant and emulator, respectively.

$$E = \frac{1}{2}(y - \hat{y})^2. \tag{5.2.11}$$

The back-error-signals for the emulator is similar to the off-line learning scheme as only one network is involved. The on-line learning computation of the neuro-control scheme can be summarized as follows:

Step 1. Read the output of the process y and compare with the desired reference r.

Step 2. The error signal δ_k^E between the output and hidden layers of the neuro-emulator is computed.

Step 3. The error signal δ_j^E between the hidden and input layers of the neuro-emulator is computed.

Step 4. The error signal δ_k^C between the output and hidden layers of the neuro-controller is computed.

Step 5. The error signal δ_j^C between the hidden and input layers of the neuro-controller is computed.

Step 6. Adapt the connection weights between the output and hidden layers and hidden and input layers, respectively, of the neuro-controller.

Step 7. Compare the output of neuro-emulator $\hat{y}$ and the actual plant output y.

Step 8. Adapt the connection weights between the output and hidden layers and hidden and input layers, respectively, of the neuro-emulator.

5.2.4 Fuzzy Logic Control Scheme

Fuzzy logic control systems are developed along linguistic lines which have some close associations with the field of artificial intelligence. Based on Zadeh's theory of fuzzy sets [23], [24], the concept of fuzzy logic has been applied successfully to the control of industrial processes particularly those which are ill-defined and those capable of successful human-operator regulation [25], [26].

A simple fuzzy logic control system as shown in Fig. 3.7.2, originally conceived by Assilian and Mamdani [27], consists of a set of linguistic conditional statements or fuzzy associations which define individual control situations. In many applications, it is sufficient to relate the error and the change in error of the process to apply the correct amount of change in the control input to satisfactorily control the system. In this respect simple linguistic rules can be formulated based on observation or a simple study about the process. The basic concept of fuzzy logic control has been discussed in Section 3.7.

For the water bath system, we specify the input variables as the performance error, e(t) where e(t) is the error between the desired temperature and the actual temperature of the water, and the rate of change of the performance error, c(t). The output or the control fuzzy variable is the voltage signal to the heater, u(t), where the range of the voltage is between 0V to 5V as described.

Change in Error, $\Delta e = c$ \ Error, e	NL	NM	NS	ZE	PS	PM	PL
PL				4 PL	3 PL	2 PL	1 PL
PM				10 PM	9 PM	8 PM	5 PL
PS			17 PS	14 PS	13 PS	11 PM	6 PL
ZE	23 NL	21 NM	18 NS	16 ZE	15 PS	12 PM	7 PL
NS			25 NS	20 NS	19 NS		
NM				22 NM			
NL				24 NL			

Fig. 5.2.3. Fuzzy rules bank matrix for controlling the water bath system.

We quantized the three fuzzy variables into seven fuzzy subsets from negative large (NL) to positive large (PL). In practice, these quantizations can be expanded according to the complexity of the control problem. For clarity the fuzzy associations or rules can be specified in matrix form as shown in Fig. 5.2.3. Basically, by using common sense and engineering judgment the water bath fuzzy association matrix can be developed, where in this case, a set of 25 rules has been formulated.

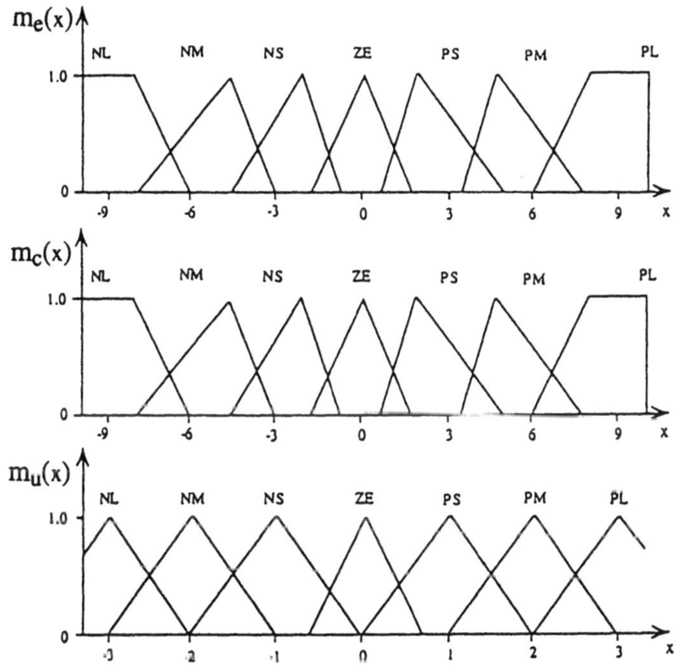

Fig. 5.2.4. Membership function of the fuzzy variables of the water bath system.

Each bank in the matrix constitutes one fuzzy association or rule i which describes how to modify the control variable u_i for observed values of the water bath input fuzzy variables e_i and c_i. As an example, we can interpret rule no. 7 (PL, ZE; PL) in natural language as follows:

If the error in the water temperature is positive and large, and the rate of change of the error is about zero, then the heater should be on positive and large.

Note that it is not necessary to fill up all the banks in the matrix. In general, some rules can be omitted and added according to the complexity of the control problem.

Triangular and trapezoidal membership functions have been chosen for the fuzzy subsets of the three fuzzy variables as shown in Fig. 5.2.4. It can be observed that the ZE (zero) fuzzy subset of the three fuzzy variables are narrower to permit finer control near the set points. Using heuristic rules, the contiguous fuzzy subsets in each library overlapped about 25 per cent. Too much overlap blurs the distinction between the fuzzy-set values. Too little overlap tends to resemble bivalent control, producing excessive overshoot and undershoot. In reality, the overlap helps smooth the transition of the control action during the operation of the controller.

The antecedents are combined by the conjunctive AND which activates the consequent fuzzy set u_i to degree w_i, therefore, using Zadeh's fuzzy 'intersection' association [23], we have

$$w_i = m_{e_i}(e) \wedge m_{c_i}(c) \tag{5.2.12}$$

where m_{e_i} and m_{c_i} are the corresponding membership values of the error and change in error.

The output fuzzy set's shape depends on the FAM-rule encoding scheme used. We used the correlation-minimum encoding scheme where the consequent fuzzy set u_i in the library of the output fuzzy set values is clipped to degree w_i with pointwise minimum with

$$m_{O_i}(y) = w_i \wedge m_{u_i}(y) \tag{5.2.13}$$

where $m_{O_i}(y)$ is the degree of membership of the activated output fuzzy set at interval y in the output universe of discourse.

In practice, $m_{O_i}(y)$ is not unique, the system then combines all of these functions as their union to yield the actual output fuzzy membership function, $m_O(y)$, following Zadeh's fuzzy 'union' association [23] such that

$$m_O(y) = w_{O_i}(y) \vee m_{O_1}(y) \vee \ldots \vee m_{O_i}(y). \tag{5.2.14}$$

The discrete fuzzy centroid defuzzification procedure [27], [28] is used to defuzzify the fuzzy control output, which is the change in the actual control input, into real values as follows:

$$u = \frac{\sum_{j=1}^{p} y_j m_O(y_j)}{\sum_{j=1}^{p} m_O(y_j)} \tag{5.2.15}$$

where the output universe of discourse is discretized into p values at interval Δy, such that

$$y_p = y_{p-1} + \Delta y.$$

Like all controllers, a FLC has a number of parameters which must be chosen by the designer in prior. In the fuzzy controller, there are three scale factors, GE, GC, and GU for the process error, change in error, and the controller's output, respectively. The selection of these scale factors is akin to the selection of the PID controller parameters and the user defined polynomials of some adaptive controllers. Systematic approaches on the selection of these scale factors have been proposed by some researchers [12], [29], [30].

The computation of the fuzzy logic controller for the temperature control of the water bath system can be summarized as follows:

Step 1. Read the output of the process y(t).

Step 2. Calculate the error e(t) and rate of change of error c(t) of the process at instant t.

Step 3. e(t) and c(t) are quantized into the appropriate fuzzy subsets.

Step 4. The membership values m_{e_i} and m_{c_i} of the error and rate of change of error, respectively, are computed from the membership functions.

Step 5. The consequent membership value m_{u_i} is activated from the rule database by combining the antecedents m_{e_i} and m_{c_i} using conjunctive AND.

Step 6. m_{u_i} is clipped to m_{o_i} using correlation minimum encoding technique.

Step 7. Actual output membership value m_o is computed by combining m_{o_i} of all other consequents activated using disjunctive OR.

Step 8. The actual output value u is computed using the discrete fuzzy centroid defuzzification procedure.

5.2.5 Generalized Predictive Control Scheme

We can summarized the computation of the GPC algorithm for the water bath temperature control system as follows:

Step 1. Read the output of the process y(t).

Step 2. Perform recursive least square estimates of the process using the data and parameter vectors in (3.6.16) and (3.6.17).

Step 3. Calculate the process step response from the estimated parameters to formulate the G matrix using

$$g_j = -\sum_{i=1}^{j} a_i g_{j-1} + \sum_{i=0}^{j-1} b_i \quad j=1, 2,..., N2. \tag{5.2.16}$$

Step 4. Vector $\mathbf{r}$ is formed from a pre-specified set-point which is formulated so as to make the current output y(t) approaches the set-point smoothly. The vector can be obtained from

$$r(t)=y(t) \tag{5.2.17}$$

$$\mathbf{r}(t+j) = \alpha r(t+j-1) + (1-\alpha) r(t+1) \quad j = 1,2,...,N2. \tag{5.2.18}$$

Step 5. Vector $\mathbf{f}$ is calculated from the data obtained at time t and is given by

$$\mathbf{f}(t+j) = F_j y(t) + z^{j-1}\left[G_j(z^{-1}) - g_0 - g_1 z^{-1} - ... - g_{j-1} z^{-(j-1)}\right] \Delta u(t). \tag{5.2.19}$$

Step 6. The control input is calculated using (3.6.15) in which only the first element of U is applied to the process.

Step 7. Data are shifted for the next sampling instants.

5.2.6 Experimental Results and Discussions

For the conventional feedback water bath control system (CFC), the following velocity-form discrete-PI algorithm was used

$$\Delta u(t) = k_c[e(t) - e(t-1)] + \frac{T}{2T_i}[e(t) + e(t-1)] \tag{5.2.20}$$

where $\Delta u(t)$ is the increment of the control input, e(t) is the performance error, k_c is the controller gain or proportional band (PB), and T_i is the integral or reset time (see Section 3.2). The tuning method of Takahashi *et al.* [31] was used in the experiments to tune the PI-controller.

For each algorithm, four groups of experiments were conducted on the water bath temperature control system. Each experiment was conducted over 100 samples using a sampling time of 30 sec. which resulted in a 50 min. duration. In order to get the best results from each experiment, heuristic corrections and fine adjustments of the relevant parameters in each scheme were made. The respective methodology and the best experimental results of each algorithm are compared and tabulated where their performances are graded comparatively.

In the first set of experiments, the tracking performances of the four controllers with respect to set-point changes were studied. In the case of the GPC, its best performance was obtained when the design parameters were selected as follows: NU=1, N1=1, N2=3, and λ=0.1. Recursive least squares estimation with no forgetting factor and a covariance matrix of diag {1,0,0} were used to estimate the system parameters in all the experiments involving the GPC. The CFC's best performance was observed when the PI-controller parameters were fine-tuned to 2.3 and 80 seconds for k_c and T_i, respectively. For the FLC, after some trial and error, the system's best performance was observed when the scaling factors were selected as follows: GE=0.16, GC=0.83, and GU=4.20. The FLC's performance was also affected when wrong consequents were chosen in the rules database, especially, those around the steady-state FAM rule. In the case of the NNC, no changes were made during the on-line learning, where the same model and learning parameters were used as in the off-line training case. It was also observed that even without the on-line learning through the emulator, the NNC performed just as well.

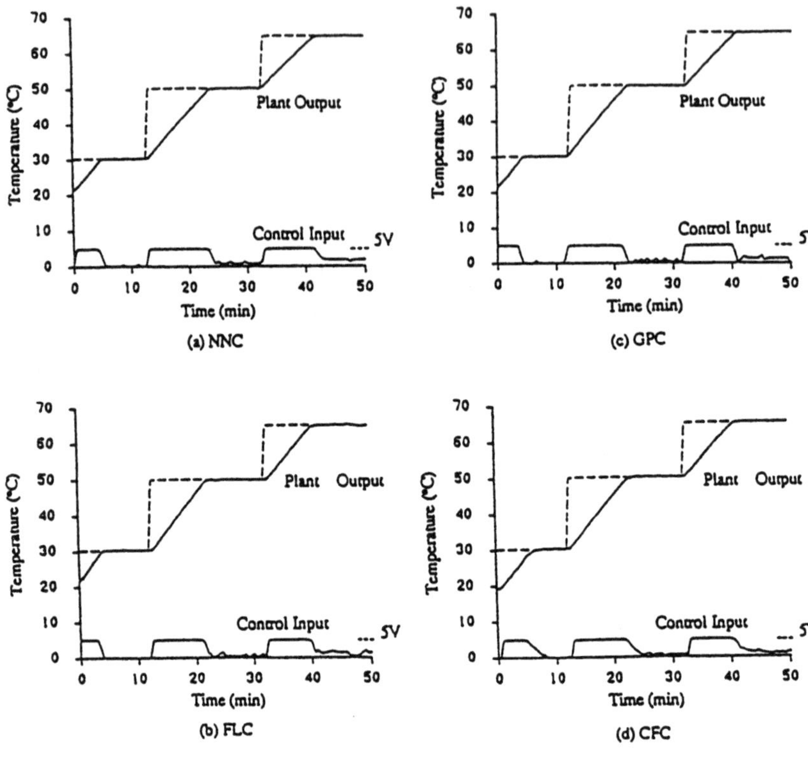

Fig. 5.2.5. Set-point tracking performance of the four controllers on the water bath system.

The four controllers performed considerably well in tracking the three given set-points and the results of the experiments are shown as in Fig. 5.2.5 (a)-(d). As expected, the CFC performed well for one operating point, i.e., in this case the upper set-point but deteriorates at another (lower set-point). The NNC shows the best tracking performance and once trained it can be applied directly without the necessity and hassle of selecting any design and tuning parameters.

The second set of experiments was carried out with the purpose of studying the ability of the four controllers in rejecting unwanted load disturbances. In order to make identical comparisons among the controllers, impulse load disturbances of values 4.0 °C and 2.0 °C were added to the process output at the 53rd and 75th sampling instants, respectively. A set-point of 50 °C was given in all the experiments.

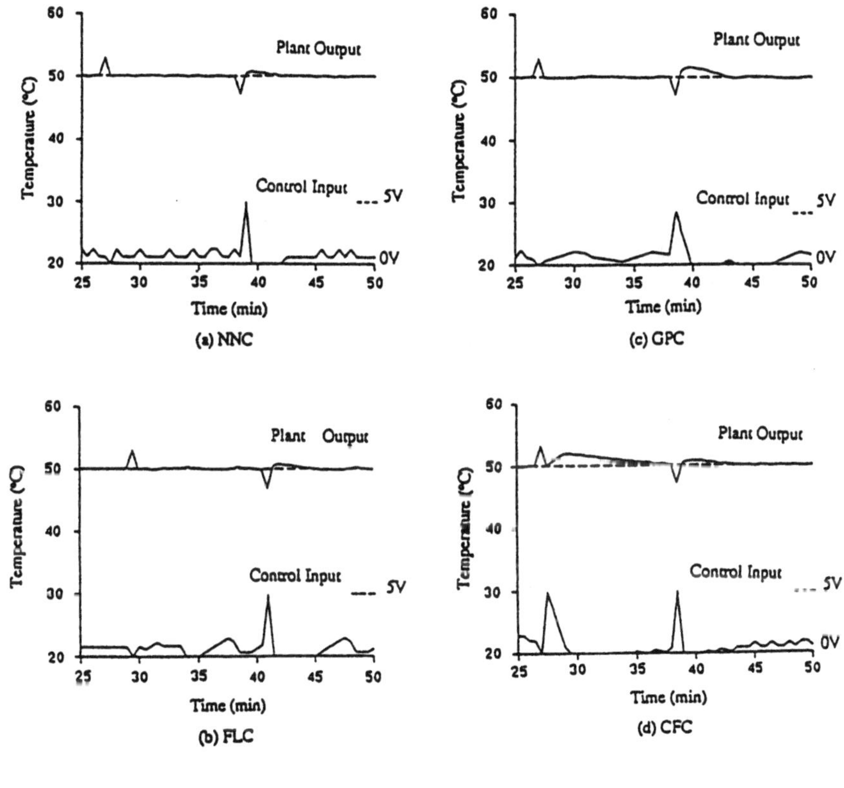

Fig. 5.2.6. Performance of the four controllers under the influence of unexpected load disturbances.

The same design and controller parameters were used for all the controllers as in the first set of experiments. The effects of the load disturbances on the four controllers are shown as in Fig. 5.2.6 (a)-(d). For clarity, only the performances of the system at the sampling period of interest are shown in the figures. It can be observed that the NNC and the

GPC performed just as well with the ability to recover very quickly from the effect of the load disturbances.

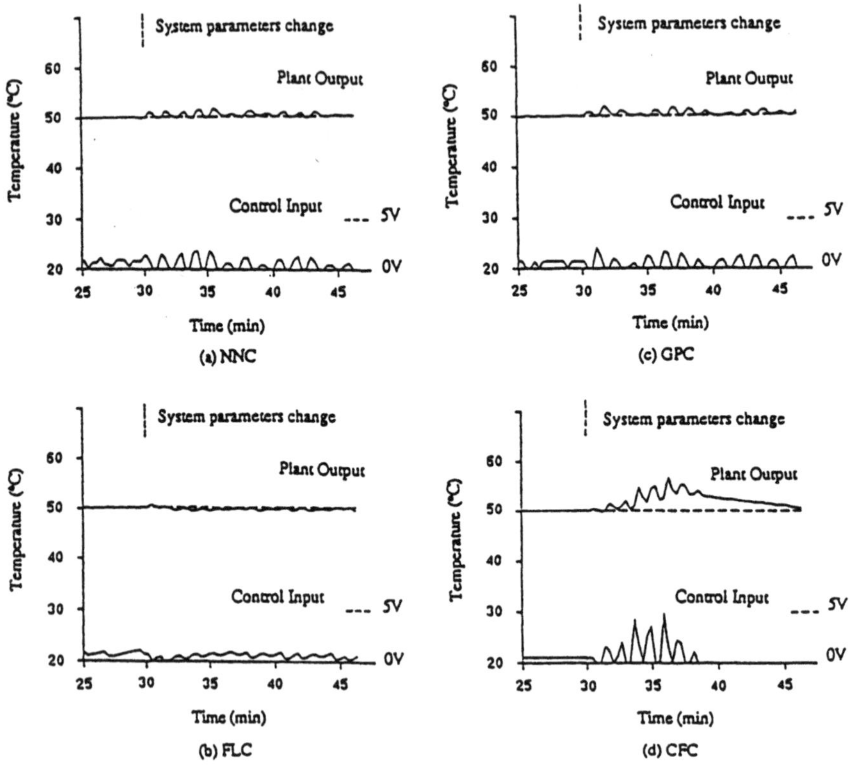

Fig. 5.2.7. Performance of the four controllers when there is a large change in the plant dynamics.

However, the system was badly affected when using the FLC and the CFC. The performance of the FLC was affected after the first disturbance, as badly as the CFC, as it happened drastically and unpredictably. In this respect, a rule-adaptive fuzzy system is more desirable as such unpredictable disturbance cannot be incorporated into the fixed fuzzy control rules.

The NNC was able to recover quickly due to its powerful generalization capability which allows it to adapt very fast to adverse changes at its inputs. The GPC was able to perform just as well as it has a j-step ahead predictor. Among the four controllers, the CFC performed worst as it is rather slow and cannot adjust quickly and automatically in rejecting unpredictable load disturbances.

One inherent characteristic of many process control systems is that their parameters tend to change in an unpredictable way. To test the robustness of the controllers, a value of 0.8u(t-2) was added to the process output after the 60th sample in each of the experiments, which represents a change in the process parameters. It is seen that these changes in dynamics are large and although it is difficult to imagine a real plant behaving in such a drastic way in practice, the experiments were conducted to illustrate the relative robustness and adaptivity of the algorithms. The behaviors of the four controllers under this influence are shown in Fig. 5.2.7(a)-(d).

Among the four controllers, as expected, the CFC behaved worst. Despite the integral action, offset is seen at the plant output due to persistent control saturation. However, the integral action did help in removing the offset eventually. The other three controllers showed considerable robustness where they were still able to track the set-point even when there is a large change in the plant dynamics. The NNC performed considerably well in tracking the set-point where its generalization capability and on-line learning help to improve its performance. The long range prediction of the GPC helps to stabilize its performance when there are parameter variations in the plant. The FLC shows the best performance among the four controllers. Unlike the load disturbance, this change in plant dynamics is not as drastic. The FLC is able to transform information regarding the process error and change-in-error into the appropriate control actions which resulted in little fluctuations at its output.

One of the common problems in controlling industrial processes is the case of processes with long and variable time-delays. Therefore, it is appropriate to compare the four algorithms in terms of robustness in dealing with a plant having such time-delays. In the fourth set of experiments, artificial time-delays of 1 sample between $0 \leq tT \leq 30$, and 2 samples between $30<tT\leq100$ were introduced in the control loop in addition to the original one sample time-delay of the plant. In these experiments two set-points were given as 35 °C between $0\leq tT\leq30$ and 55 °C between $30<tT\leq100$ and only the time-delay of the plant between $0\leq tT\leq30$ was assumed known in prior.

The neuro-controller and emulator were re-trained from the input-output information of the plant between $0\leq tT\leq30$. However, the structure of the neural network and the number of training patterns used were kept the same. In the GPC, the estimation of the $B(z^{-1})$ parameters was increased to 4 and the design parameters were changed as follows: NU=2, N1=1, N2=10, and λ =0.5 to accommodate for the time-delay of the plant. Similarly, the PI controller was re-tuned where k_c and T_i were increased to 3.0 and 100 sec., respectively. The scaling factors of the FLC were re-scaled where GE=0.11, GC=0.70, however, GU remained the same.

The performances of the four controllers on the same plant with a variable time-delay are shown as in Fig. 5.2.8 (a)-(d). As expected, the GPC performed best under this condition as the algorithm is suited for plants with variable time-delays provided that the number of estimated $B(z^{-1})$ parameters covers the range of possible time-delay. Using the NNC, overshoot occurred at the second set-point when the plant deadtime changed, however, the on-line learning helps to improve its performance eventually. The FLC performed slightly better than the NNC. The nonlinear defuzzification algorithm of the FLC help to reduce overshoots in the system. The CFC performed worst during the second set-point as its output oscillated adversely. This is because the tuning method used relies on the plant's time-delay and its open loop step response.

In all the experiments, it can also be observed that the outputs of the NNC and FLC are much smoother than the GPC and CFC. Sharp fluctuations can be observed more clearly at the output of the GPC, which resulted in minor fluctuations at the plant outputs. Prolong adverse fluctuations of the control signals could also damage the actuator. The nonlinear mapping ability of the NNC and FLC lead to smoother controller outputs which could prolong actuator life.

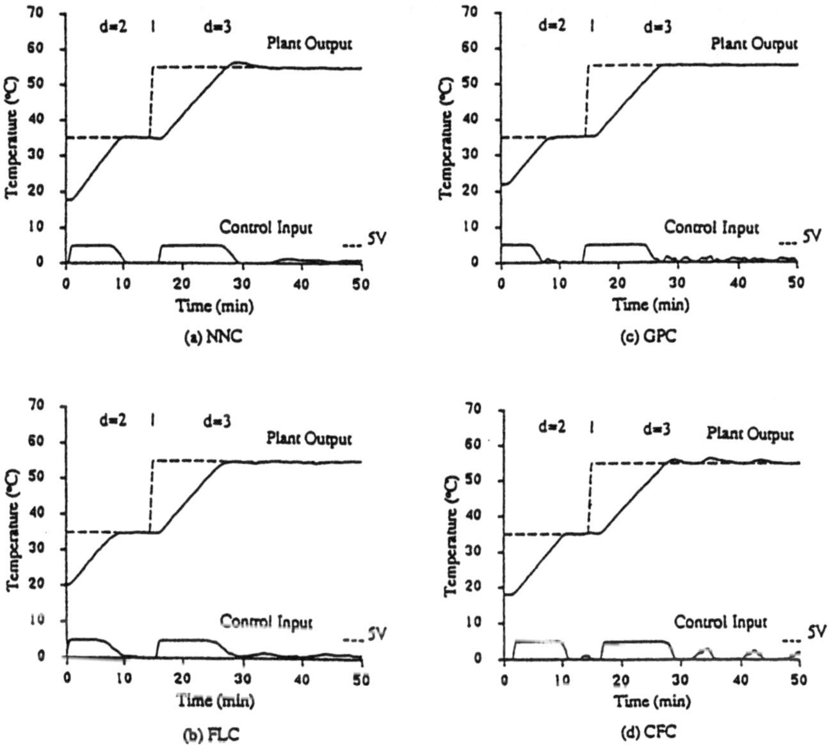

Fig. 5.2.8. Performance of the four controllers when then plant has long and variable time-delay.

5.2.7 Conclusions

This study compares four control schemes for the temperature regulation of a water bath system. The four schemes have some similarities and also dissimilarities in their approaches. From the results of the experimental studies, apart from the CFC, we found that each of the control schemes has both desirable and undesirable characteristics. Among the four control schemes, the CFC scheme is perhaps the simplest to be implemented but it was found that this conventional controller cannot cope with complexities in the process. For simple and non-demanding control problems the CFC is, thus, recommended.

As the reduced-order mathematical model of the water bath can be obtained, this information is used to develop the four controllers. At first, rough tuning of both the FLC and CFC parameters were done in simulation and these values were further fine-tuned to give the best performance on-line. For the NNC, the number of input vector elements of the neural network models follows the reduced order model of the plant. However, for applications where the plant models are ill-defined [27], the NNC, FLC, and CFC can still be applied without the need for any mathematical models. In this respect, without *a priori* mathematical models, the GPC and many traditional adaptive controllers are impossible to be developed.

One drawback in the neuro-control approach is the prior training. A judicious choice of the model structure and learning parameters is also essential for the success of the neural networks. However, once trained, the NNC is self-tuned and does not consist any requirement for tuning. In the case of the GPC even though it is self-tuned, there is some difficulty in selecting the user-defined parameters, especially, if the plant is complex.

For the FLC, there are a number of factors and values that need to be chosen in prior to ensure the success of the controller which are quite laborious. Although some of these values may be determined heuristically or by trial and error, for example, the width of the membership functions, scaling factors, etc., fine scaling of these values is essential to obtain the best FLC's performance. In this respect the adaptive fuzzy systems of [11], [12], [13] are more desirable where in addition the rules can also be generated automatically. However, these adaptive fuzzy systems may be more complex to be developed. The CFC has to be tuned manually and occasionally if the operating point of the process changes due to disturbances and other factors.

Another advantage of neural network and fuzzy logic control over traditional adaptive and conventional control is that they can be used to solve highly nonlinear control problems. There has been many reports published where neural networks can be used to control highly complex nonlinear control systems [8], [9]. The semilinear sigmoidal functions in the hidden neurons allow the network to learn and solve nonlinear control problems where to this end traditional control approaches have no solution yet. It has also been reported that the FLC too can be used for such purpose [6], [11], [13]. Sources of nonlinearity in the FLC include the fuzzification and defuzzification algorithms, the FAM rules and logic inference.

Table 5.2.1. A summary of comparisons among the four control schemes based on the experimental results.

CRITERIA	NNC	FLC	GPC	CFC
Plant Math Model	Not Necessary	Not Necessary	Necessary	Not Necessary
Tuning of Controller Parameters	Self-tuned, no parameters needed to be tuned, but neural networks has to be chosen and trained in prior	Self-tuned but scaling factors have to be , GE, GC, and GU have to be selected in prior	Self-tuned, but control horizons, N1, N2, and NU and control weighting λ have to be selected in prior	Controller parameters K_P, K_I, and K_D (or k_c, T_i, and T_d) have to be tuned
Computational Complexity	Heavy	Lightest	Heaviest	Light
Tracking Performance	Best	Worse	Medium	Worst
Disturbance Rejection	Best	Worse	Medium	Worst
Plant Parameter Variations	Medium	Best	Medium	Worst
Variable Time-Delay	Worse	Medium	Best	Worst
Smoothed Control Signal	Medium	Medium	Worst	Worse

The neural network and fuzzy logic control algorithms can be feasibly implemented in parallel hardwares for high speed operation [3], [4], [5]. It has been reported in [4] that the speed of inference in a custom VLSI fuzzy chip is independent of the number of rules. Moreover, the parallel processing operation of these controllers are more robust even if some of the synaptic connections failed [6], [32]. The GPC's computation time is very slow among the four algorithms as it involves a lot of calculations. The FLC's computation time is the lightest as it mainly involves logic operations and comparison. A summary of the experimental results is shown in Table 5.2.1.

This study has shown that artificial neural networks can be feasibly implemented for real-time control implementation. The neuro-control approach is robust and its performance is comparable to some established traditional control methods. Although the neuro-control approach exhibits desirable characteristics and offers encouraging advantages not found in the other three controllers, it is hoped that by extracting the best characteristics of each class of systems in this study, a more efficient and reliable control scheme can be developed.

5.3 Stabilizing an Inverted Pendulum by Neural Networks

5.3.1 Introduction

Rapid progress in micro-electronics technology enables us to use intelligent control methods. Though neuro-control has yet to make an impact, fuzzy control and other traditional control approaches have been well-developed and applied to many real control problems quite satisfactorily. In this application, we consider the stabilization problem of an inverted pendulum. An inverted pendulum is an example of an unstable control system and hence, posing a challenging problem for control engineers. The inverted pendulum considered here is attached to the top of a cart which can be controlled to move to the right or left in the horizontal direction. The control problem is to stabilize the inverted pendulum in a vertical position as long as possible.

First, a fuzzy controller is designed to swing up the inverted pendulum from the bottom and stabilize it at a vertical position. We use two sets of fuzzy rules for swinging up and stabilizing the pendulum. To swing up the pendulum, we use the first set of seven fuzzy rules. When the pendulum angle, i.e., its angular velocity and the cart position are in some specified regions, the fuzzy controller is switched to the second set of fuzzy rules. The second set consists of eleven fuzzy rules for stabilizing the pendulum, however, it is rather difficult to select the best set of rules for good control.

In order to control it more precisely we apply the technique of optimal control in the range where linear approximation can be used in describing the linearized model of the inverted pendulum. Finally, we apply a neural network to adjust the discrepancy between the pendulum position and its desired reference. The performance of the linear quadratic optimal controller can be further improved by adding a neural network which compensates for nonlinearities such as backlash and friction in the system. This can be achieved by using a neural network with a small number of hidden neurons. Experiments are conducted on the inverted pendulum in real-time using these control laws.

5.3.2 Description of the Inverted Pendulum System

The inverted pendulum system consists of three parts; (i) the pendulum rod on a cart, (ii) a control box, and (iii) a personal computer with an A/D and D/A converters. A schematic diagram of the inverted pendulum control system is illustrated in Fig. 5.3.1 and the control flow chart is given in Fig. 5.3.2. The mechanical inverted pendulum and the control box have been developed by Advanced Control Laboratory Inc. of Japan and the personal computer is an NEC PC-9801RA with a math-co-processor. Table 5.3.1 gives the physical parameters of the inverted pendulum.

The outputs of the inverted pendulum system are the pendulum angle θ and the cart position x. The pendulum angle θ is measured from a vertical line and the cart position x is measured from the center of the cart track as shown in Fig. 5.3.1. The input to the system is a horizontal force applied to the cart.

The control objective is to move the cart at position $x = 0$ in such a way that the pendulum continues to stand in the vertical line. It means that $\theta=\dot{\theta}=0$ at $x = 0$ where $\dot{\theta}$ denotes derivative of θ with respect to time. To stand the inverted pendulum upright from its initial stable position downwards, it has to be swung up by giving a force to the cart to move in

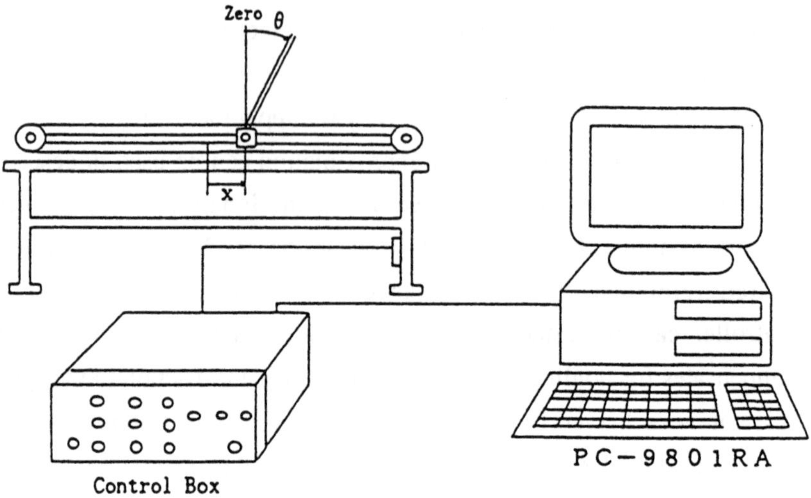

Fig.5.3.1. Inverted pendulum control system.

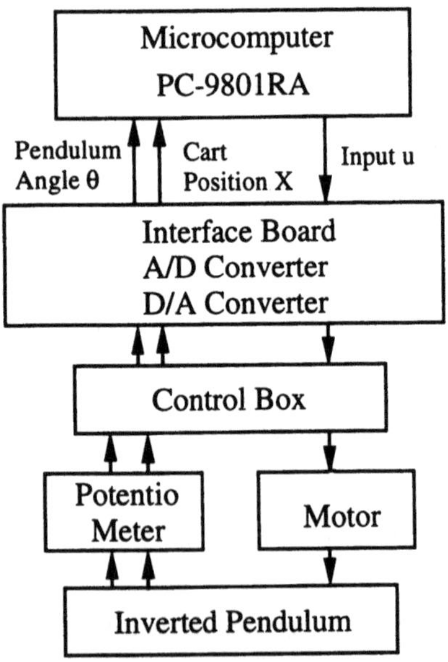

Fig.5.3.2. Control flow chart of the inverted pendulum.

the horizontal line. This is referred to as 'swing-up control'. Then the inverted pendulum is to be stabilized vertically in an upright position by controlling the cart in the horitzontal x direction which we refer to as 'stabilization control'.

5.3.3 Initial Start-Up Control Using Fuzzy Logic

As discussed we first consider the 'swing-up control' by a set of seven fuzzy rules as shown in Table 5.3.2. In order to swing up the pendulum from its stable vertically downwards position, it is necessary to use its kinetic energy efficiently. The signs of the co-ordinates to move the cart or to stand the pendulum upright are shown in Fig. 5.3.3. The membership function chosen to control the pendulum is shown in Fig. 5.3.4.

From its stable vertically downwards position, the inverted pendulum is swung up by moving the cart towards the right or left direction in the horizontal line. Since θ is ZO and $\Delta\theta$ is ZO at the beginning, from Table 5.3.2 the control action for the cart is PM where $\Delta\theta$ denotes the difference of θ, which is the approximated value of $\dot{\theta}$.

Table 5.3.1. Physical parameters of the inverted pendulum.

Items	Specifications
Equivalent mass of the cart	4.4kg
Voltage/torque conversion coefficient	20.09N/V
Equivalence friction coefficient of the cart	19.2 Kg/s
Moment of the center of gravity of pendulum	2.14×10^{-3} Kgm^2
Viscosity friction coefficient of axis	0.115 m
Distance from center of gravity of pendulum	4.9×10^{-4}Kgm^2/s
Mass of the pendulum	0.1 Kg
Position/voltage conversion coefficient	1.7V/m
Angle/voltage conversion coefficient	3.34V/rev

Applying the fuzzy rules, the cart will move towards the positive direction. When the cart is moved towards the negative direction, the potential energy of the pendulum increases. Then the pendulum begins to move downwards such that it decreases the potential energy and increases the kinetic energy. Thus, the pendulum will swing down again and at the bottom θ is ZO and $\Delta\theta$ is NM.

From Table 5.3.2 the next control action is NM. Repeating this procedure step by step, the inverted pendulum swings up and down until θ becomes larger. If the inverted pendulum enters into a region where the pendulum angle θ and the angular velocity $\dot{\theta}$ are preassigned, we switch the control action from 'swing-up' control to 'stabilization control'. This region is specified as $\theta \in [-3, 3]$ and $\Delta\theta \in [-2, 2]$. However, if $x \in [-5, 5]$ then the cart moves toward the origin and the 'swing-up control' is repeated. This region is referred to as the 'stable region'.

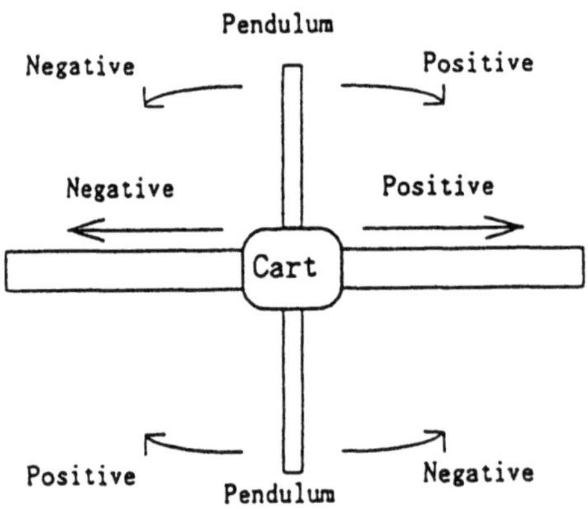

Fig. 5.3.3. Direction signs of the inverted pendulum system.

Next, we consider 'stabilization control' when the inverted pendulum has entered into the 'stable region'. The fuzzy rules for the 'stabilization control' part are shown in Table 5.3.3 which can be interpreted as follows:

Rule 1: If θ = PM and $\Delta\theta$ = ZO, then v = PM.

Rule 2: If θ = PS and $\Delta\theta$ = PS, then v = PS.

Rule 3: If θ = PS and $\Delta\theta$ = NS, then v = ZO.

Rule 4: If θ = NM and $\Delta\theta$ = ZO, then v = NM.

Rule 5: If θ = NS and $\Delta\theta$ = NS, then v = NS.

Rule 6: If θ = NS and $\Delta\theta$ = PS, then v = ZO.

Rule 7: If θ = ZO and $\Delta\theta$ = ZO, then v = ZO.

Rule 8: If θ = ZO and $\Delta\theta$ = NM, then v = NM.

Rule 9: If θ = ZO and $\Delta\theta$ = NS, then v = NS.

Rule10: If θ = ZO and $\Delta\theta$ = PS, then v = PS.

Rule 11: If θ = ZO and $\Delta\theta$ = PM, then v = PM.

Here, v denotes input voltage to the motor of the cart.

Table 5.3.2. Fuzzy rules for swinging-up the pendulum.

θ \ $\Delta\theta$	NM	NS	ZO	PS	PM
NM	ZO				
NS	NS				
ZO	NM		PM		PM
PS					PS
PM					ZO

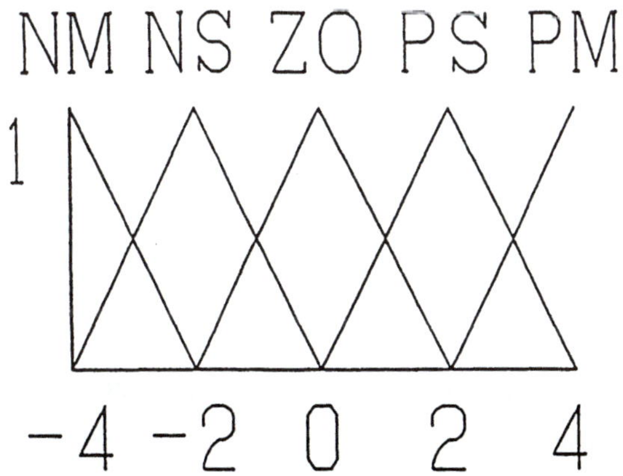

Fig. 5.3.4. Membership function assigned for all the fuzzy variables of the inverted pendulum system.

Table 5.3.3. Fuzzy rules for the stabilization of the pendulum.

θ \ Δθ	NM	NS	ZO	PS	PM
NM			NM		
NS		NS		ZO	
ZO	NM	NS	ZO	PS	PM
PS		ZO		PS	
PM			PM		

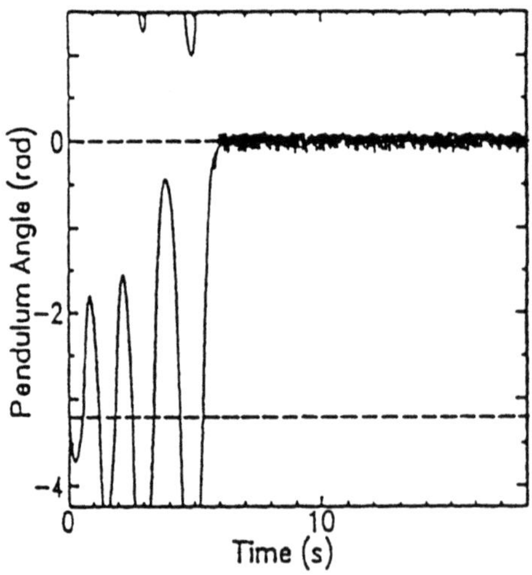

Fig. 5.3.5. Experimental results to stabilize the pendulum using the fuzzy rules of Table 5.3.3.

The fuzzy controller is based on the min-max center of gravity method by Assilian and Mamdani [27] given by (5.2.15). Applying this control action, the inverted pendulum can stand upright when the cart moves towards the right and left directions very rapidly. Fig. 5.3.5 shows the experimental results and it can be observed that even though the inverted pendulum stands upright all the time the fluctuation seems to be rather large using only the fuzzy logic controller with the set of rules in Table 5.3.3.

5.3.4 Using Optimal Control Strategy for Stabilization of the Pendulum

The fuzzy logic controller is basically used as an initial start-up for the inverted pendulum. The performance of the system can be improved by adding more fuzzy rules in the fuzzy matrix bank. However, due to limitation of memory capacity and speed of the personal computer, we are unable to increase the fuzzy rules for such a fast real-time application. Moreover, it is difficult to select the best set of rules to improve the performance of the fuzzy controller. However, the second set of fuzzy rules which is used for the stabilization of the pendulum is mainly used for the purpose of coarse control.

We note that in the 'stable region' the nonlinear dynamics describing the inverted pendulum can be approximated by a linear dynamical system. In this case we can apply traditional control methods to stabilize the inverted pendulum. In this application, linear quadratic optimal control theory [35] is used. First, we consider the equations of motion of the inverted pendulum and the cart such that

$$(M+m)\ddot{x} + ml\ddot{\theta}\cos\theta - ml\dot{\theta}^2 \sin\theta - u + F\dot{x} = 0 \quad (5.3.1)$$

$$(J+ml^2)\ddot{\theta} + ml\ddot{x}\cos\theta - mgl\sin\theta + C\dot{\theta} = 0 \quad (5.3.2)$$

where M is the equivalent mass of the cart, m is the mass of the inverted pendulum, F is equivalent friction coefficient of the cart, l is distance from axis to the center of gravity of the pendulum, J is moment of inertia of the center of gravity for the pendulum, C is viscosity friction coefficient of axis, and g is acceleration of gravity.

If we can approximate $\cos\theta \approx 1, \theta^2\dot{\theta} \approx 0,$ and $\sin\theta \approx 0$, then the nonlinear system dynamics of (5.3.1) and (5.3.2) become linear. Then we have

$$\dot{x}(t) = Ax(t) + bu(t) \tag{5.3.3}$$

$$u(t) = cx(t) \tag{5.3.4}$$

where

$$A = \begin{bmatrix} 0 & 1 & 0 & 0 \\ 0 & \frac{-F(J+ml^2)}{D} & \frac{-m^2gl^2}{D} & \frac{mlC}{D} \\ 0 & 0 & 0 & 1 \\ 0 & \frac{mlF}{D} & \frac{(M+m)mgl}{D} & \frac{(M+m)C}{D} \end{bmatrix}$$

$$b = \begin{bmatrix} 0 & \frac{J+ml^2}{D} & 0 & \frac{1}{D} \end{bmatrix}^T$$

$$c = \begin{bmatrix} 1 & 0 & 0 & 0 \\ 0 & 0 & 1 & 0 \end{bmatrix}$$

$$D = (M+m)(J+ml^2) - m^2l^2$$

$$x = [x_1, x_2, \theta_1, \theta_2]^T$$
$$x_1 = x,\ x_2 = \dot{x}_1$$
$$\theta_1 = \theta,\ \theta_2 = \dot{\theta}_1. \tag{5.3.5}$$

The discrete-time representation of the above equations are as follows:

$$x(t+1) = Px(t) + qu(t)$$
$$y(t) = cx(t)$$
$$P = e^{AT},\quad q = Q_0 b$$
$$Q_0 = \int_0^T e^{A\tau} d\tau \tag{5.3.6}$$

where T denotes the sampling interval. The optimal control strategy is to minimize the cost function J given by

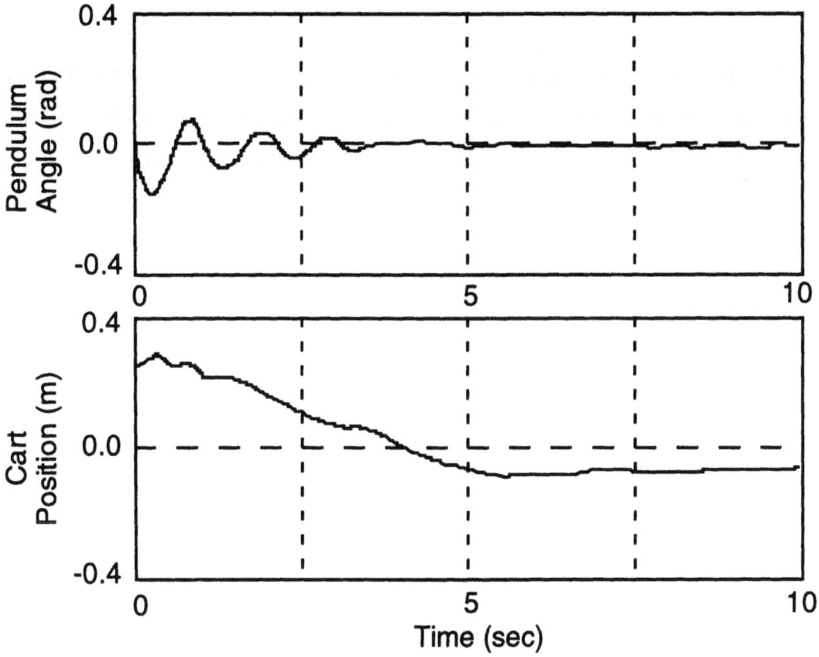

Fig. 5.3.6. Experimental result by using linear quadratic optimal control with the minimal order observer.

$$J = \sum_{t=0}^{\infty} (y(t+1)^T y(t+1) + \Gamma u^2(t)), \quad \Gamma > 0. \tag{5.3.7}$$

The optimal control signal u(t) is

$$u(t) = -Kx(t)$$

$$K = \left(\Gamma + q^T H q\right)^{-1} q^T H P \tag{5.3.8}$$

where H denotes a steady solution of the Riccati equation given by

$$\begin{aligned} H(t+1) &= P^T H(t) P \\ &\quad - PH(t)q(\Gamma + q^T H(t) q)^{-1} q^T H(t) P + c^T c \end{aligned}$$

$$H(0) = \alpha I, \quad \alpha > 0. \tag{5.3.9}$$

Using the above linear approximation of the system, we can directly apply the linear quadratic optimal control theory for stabilizing the pendulum. Since x(t) is unknown, we must estimate it from observation data $y(n),\ n \le t$. Here, we develop an observer where the estimate of x(t) is denoted by $\hat{x}(t)$. The observer can be constructed such that $\hat{x}(t)$ becomes x(t) when t is sufficiently large. The following feedback law is applied

$$u(t) = -K\hat{x}(t). \tag{5.3.10}$$

Figure 5.3.6 shows the experimental results by using the optimal control law as described above. The results show further improvement in the performance of the inverted pendulum.

5.3.5. Fine Improvement by Neural Networks

In the previous section we consider the control method which combines fuzzy and optimal control theories. Even if the weighting function in the

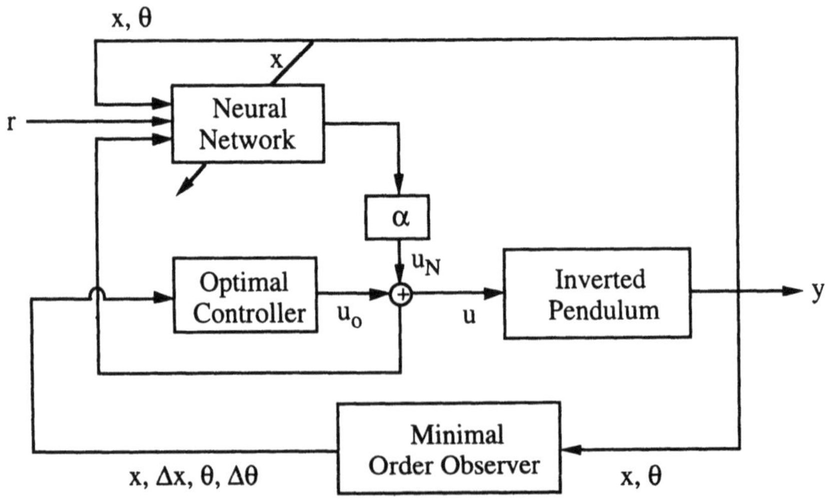

Fig. 5.3.7. Inverted pendulum system with neural network compensator.

cost function J is changed, it was impossible to remove the offset of the position x from the origin. This is due to the nonlinearity of the friction coefficient of the system and hence, the small power of the motor could not overcome such a problem.

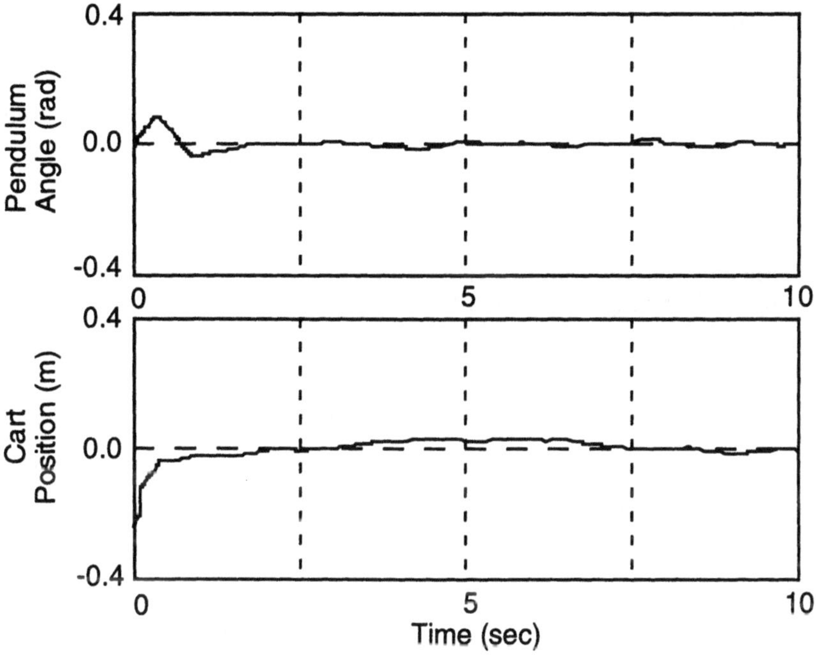

Fig. 5.3.8. Experimental results of the inverted pendulum system compensated by the neuro-controller.

Neural networks can represent nonlinear input/output mapping and realize a nonlinear controller. Thus, we adopt the neuro-control strategy to improve the control performance. The configuration of the present neuro-controller is illustrated in Fig. 5.3.7 where α is a positive small number, NN denotes a layered neural network, and r(t) is a reference value. In Fig. 5.3.7, u_0 is the optimal control and u_N denotes the output of the neuro-controller which minimizes x^2. Using this control scheme, we can remove the offset as shown in Fig. 5.3.8.

5.3.6 Conclusions

In this case study we have applied a hybrid method of traditional and advanced control strategies to control an inverted pendulum which is an example of an unstable system. Using a set of fuzzy rules, the inverted pendulum can be swung up from its vertically downwards stable position. The pendulum was initially kept stable in the upright position by another set of fuzzy rules which provided some sort of coarse control. Once it was in the upright position, the control of the pendulum was taken over by linear optimal control strategy where its dynamical model was linearized. A neural network having a small number of hidden neurons was finally used to compensate offset errors due to nonlinearities in the system.

5.4 Speed Control of an Electric Vehicle by Self-Tuning PID Neuro-Controller

5.4.1 Introduction

Due to environmental problems the automobile industry is currently venturing into producing electric vehicles. At the Shikoku Electric Power Company, Japan, a new type of electric car which is called PIVOT has been developed. It can rotate around a center and shift laterally. To control the car effectively, a simulator system of an electric vehicle driven on real road conditions has been developed. In the electric vehicle, the running resistance can be realized by controlling the torque of a motor which is operated as a generator.

We have developed a self-tuning PID neuro-controller to control the speed and load torque of the electric vehicle. In this development the PID gains are tuned automatically by the neural network in an on-line way. Apart from the advantages of self-learning and self-organization, the neural network can be used for solving nonlinear problems as well as suppressing noise that is inherent in the system. Several experiments are conducted using the self-tuning PID neuro-control scheme. The experimental results obtained by using the proposed scheme are compared to those of a conventional PID controller. It is observed that the neural network-tuned PID controller gives better performance with reduced performance error in each case.

5.4.2 The Electric Vehicle Control System

Figure 5.4.1 shows the schematic diagram of the electric vehicle system which consists of two motors and a microcomputer. M1which is a d.c. motor operates as a generator and acts as the running load of the electric vehicle. M2 which is an a.c. induction motor is used as the driving motor which is rotated according to the scheduled speed. This configuration has been used since the torque of the d.c. motor is easily controllable and also the induction motor is light and more robust. The two motors, M1 and M2, are controlled directly by two digital controllers, C1 and C2, respectively. In this configuration, the system can simulate the real driving condition of the electric vehicle.

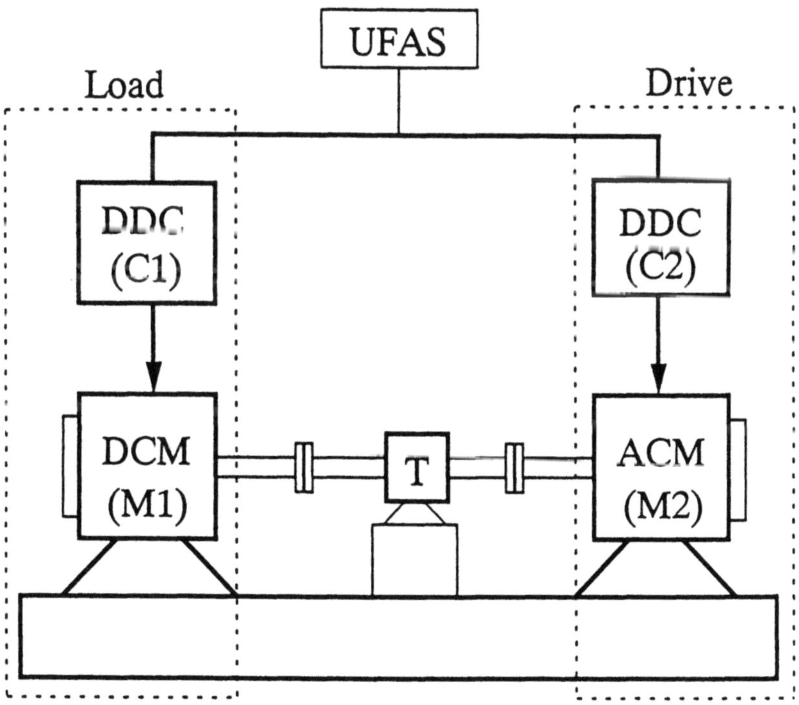

Fig. 5.4.1. Schematic diagram of the generator system of the electric vehicle. UFAS is the universal FA system, DDC is a direct digital controller, DCM is a d.c. motor, ACM is an a.c. motor, and T is a torque meter.

The output torque is controlled so that the flywheel of the output motor follows the reference speed (r(t)), which is calculated by the microcomputer (UFAS model) based on the electric vehicle running pattern. The control system has double feedback loops of speed and current. The control parameters of the conventional PID controller are fixed such that the integral time is 0.2 sec. and the proportional band (100 x inverse value of proportional gain) is 5 per cent and the derivative time is neglected. This PID controller has been tuned using the tuning methods of Takahashi *et el.* [31].

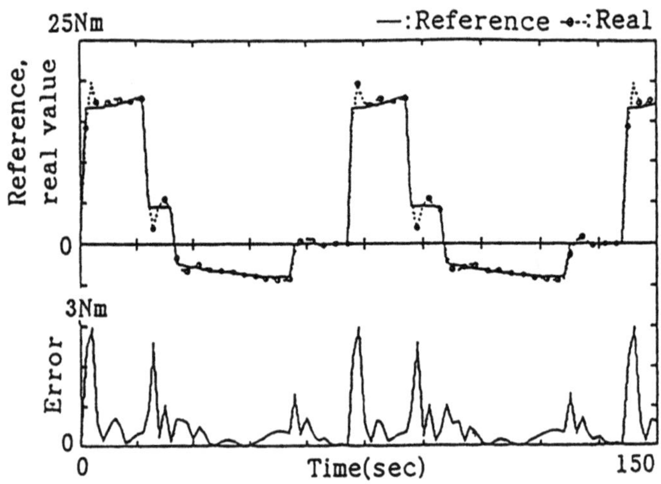

Fig. 5.4.2. Experimental results for motor torque control by conventional PID controller.

Figure 5.4.2 shows the results of the control experiment to drive the 1,000 kg electric vehicle using the conventional PID controller. The electric vehicle is driven from 0 km/h - 60 km/h, then held at constant speed for 10 sec., and is decelerated. The acceleration and deceleration rates are 4 km/h/s and 2 km/h/s. respectively. It can be observed that the error is between 1.5 and 3.0 Nm from 0 to 375 sec. In the following section, we will describe the same PID controller tuned using a neural network and then apply it for the same control problem under identical conditions.

5.4.3 Self-Tuning PID-Type Neuro-controller

Figure 5.4.3 shows the block diagram of the self-tuning PID neuro-control scheme of the electric vehicle control system. This type of neuro-controller is meaningful from a viewpoint of application because PID controllers are so widely used today. The transfer function Gc(s) of the PID controller is given as

$$G_c(S) = k_c[1 + (1/T_i)(1/s) + T_d s]. \tag{5.4.1}$$

Here, k_c is proportional gain, T_i is integral time, and T_d is derivative time. In a discrete-time system, it is possible to approximate control input as follows:

$$\begin{aligned} u(t) = {} & u(t-1) + K_P(t)[e(t) - e(t-1)] \\ & + K_I(t)e(t) + K_D(t)[e(t) - 2e(t-1) + e(t-2)] \end{aligned} \tag{5.4.2}$$

$$\begin{aligned} & K_P(t) = k_c - K_I(t)/2, K_I(t) = k_c T/T_i, \\ & K_D(t) = k_c T_d/T \end{aligned} \tag{5.4.3}$$

where u(t) is the control input, e(t) is output error, $K_p(t)$, $K_I(t)$, and $K_D(t)$ are the P, I, and D gains, respectively, and T is a sampling interval.

The neural network (NNC) adjusts the PID gains which are the 3 outputs of the NNC so as to minimize the cost function as follows:

$$J = \left(\frac{1}{2}\right)e(t)^2 \tag{5.4.4}$$

where

$$e(t) = r(t) - y(t). \tag{5.4.5}$$

The connection weight $w_{ji}(t)$ of NNC is given by

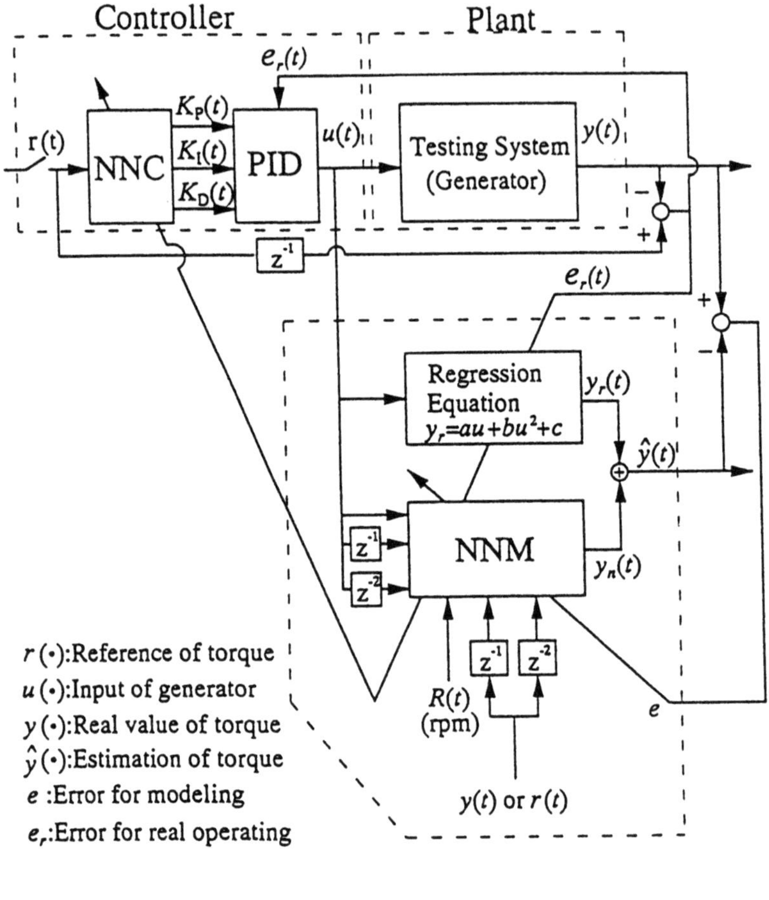

Fig. 5.4.3. Self-tuning PID neuro-control scheme of the electric vehicle speed control system.

$$\Delta w_{ji}(t) = -\eta \frac{\partial J}{\partial w_{ji}(t-1)} + \alpha \Delta w_{ji}(t-1) + \beta \Delta w_{ji}(t-2) \tag{5.4.6}$$

$$w_{ji}(t) = w_{ji}(t-1) + \Delta w_{ji}(t) \tag{5.4.7}$$

where η is the learning rate, α is the momentum rate, and β is the acceleration rate [6]. The term $\partial J/\partial w_{ji}$ (t-1) is calculated by the following equations:

$$\frac{\partial J}{\partial w_{ji}(t-1)} = -[r(t) - y(t)] \frac{\partial y(t)}{\partial w_{ji}(t-1)}$$

$$= -e(t) \frac{\partial y(t)}{\partial w_{ji}(t-1)} \tag{5.4.8}$$

$$\frac{\partial y(t)}{\partial w_{ji}(t-1)} = \frac{\partial y(t)}{\partial u(t-1)} \frac{\partial u(t-1)}{\partial w_{ji}(t-1)} \tag{5.4.9}$$

$$\frac{\partial u(t-1)}{\partial w_{ji}(t-1)} = \frac{\partial \mathbf{K}(t-1)}{\partial w_{ji}(t-1)} [\frac{\partial u(t-1)}{\partial \mathbf{K}(t-1)}]^T. \tag{5.4.10}$$

where $\mathbf{K}(t-1) = [K_P(t-1), K_I(t-1), K_D(t-1)]$. Note that $\mathbf{K}$(t-1) is an output vector of NNC, and the gradient with respect to w_{ji} (t - 1) is given by

$$\frac{\partial \mathbf{K}(t-1)}{\partial w_{ji}(t-1)} = [\frac{\partial K_P(t-1)}{\partial w_{ji}(t-1)}, \frac{\partial K_I(t-1)}{\partial w_{ji}(t-1)}, \frac{\partial K_D(t-1)}{\partial w_{ji}(t-1)}]. \tag{5.4.11}$$

Furthermore, we get from (5.4.2)

$$\frac{\partial u(t-1)}{\partial \mathbf{K}(t-1)} \equiv [u_1(t-1), u_2(t-1), u_3(t-1)]$$

where (5.4.12)

$$u_1(t) = e(t) - e(t-1)$$
$$u_2(t) = e(t)$$
$$u_3(t) = e(t) - 2e(t-1) + e(t-2).$$

Elements of (5.4.11) are obtained by the backpropagation method and $\partial J/\partial w_{ji}$ is given as follows:

$$\frac{\partial J}{\partial w_{ji}(t-1)} = -e(t)\frac{\partial y(t)}{\partial u(t-1)}[u_1(t-1)\frac{\partial K_P(t-1)}{\partial w_{ji}(t-1)} + u_2(t-1)\frac{\partial K_I(t-1)}{\partial w_{ji}(t-1)} + u_3(t-1)\frac{\partial K_D(t-1)}{\partial w_{ji}(t-1)}. \quad (5.4.13)$$

For better accuracy, another neural network (referred as NNM) is used to estimate the Jacobian $\partial y(t) / \partial u(t-1)$ of the plant.

5.4.4 Application to Speed Control of Electric Vehicle

5.4.5.1 Adjustment of Proportional Band

Here, we apply the NNC to adjust the proportional band of the PID cntroller shown in Fig. 5.4.3. Its transfer function $G_S(s)$ is as follows:

$$G_s(s) = (100 / P_B)[1 + (1 / T_I)(1 / s)] \quad (5.4.14)$$

$$P_B = P_{BF} + P_{BN} \quad (5.4.15)$$

where P_B is the proportional band, P_{BF} is its constant part (5 per cent), and P_{BN} is the parameter adjusted by NNC. T_i is the integral time (0.2 sec.) and k_c in (5.4.3) is $100/(P_{BF} + P_{BN})$. Thus, $K_P(t)$ and $K_I(t)$ can be tuned properly by adjusting P_{BN}. In this application $K_D(t)$ is not used. The control input u(t) is obtained as given in (5.4.16) derived from (5.4.2) and (5.4.3).

$$u(t) = u(t-1) + [1 / (P_{BF} + P_{BN})] [(1 - T / 2T_i)(e(t) - e(t-1)) + (T / T_i)e(t)]. \quad (5.4.16)$$

The cost function J_R is defined as follows:

$$J_R = \frac{1}{2}[R_{ref}(t-m) - R(t)]^2 \tag{5.4.17}$$

where R_{ref} (t-m) is a desired speed at time (t-m)T and R(t) is the real value. Here, mT and T are the sampling times of the microcomputer and controller, respectively, where mT is about 1 sec. and T is about 0.005 sec.. The reason for this difference between mT and T is that the NNC is being realized by a microcomputer of relatively slower speed while the DDC is realized by a special hardware device. Thus, it should be noted that the connection weights and the control signal can only be updated at every mT whereas the sampling time of the real electric vehicle system is actually T.

In this case, the output of NNC is only P_{BN} and $\partial J_R / \partial w_{ji}$ (t- m) is derived in the same way as (5.4.13) as follows:

$$\frac{\partial J_R}{\partial w_{ji}(t-m)} = -[R_{ref}(t-m) - R(t)]\frac{\partial R(t)}{\partial u(t-m)} \frac{\partial u(t-1)}{\partial P_{BN}}\frac{\partial P_{BN}}{\partial w_{ji}(t-1)} \tag{5.4.18}$$

$$\frac{\partial u(t-1)}{\partial P_{BN}} = [-1/(P_{BF}+P_{BN})^2] .[(1-T/2T_i)(e(t-1) - e(t-2)) + (T/T_i)e(t-1)]. \tag{5.4.19}$$

5.4.4.2 *Adjustment of Integral Time*

For this particular case we set $T_i=T_{if}+T_{iN}$ where T_{if} is constant (at 0.2 sec.) and T_{iN} is adjusted by the NNC. The derivative $\partial J_R / \partial w_{ji}(t-1)$ is derived as follows:

$$\frac{\partial J_R}{\partial w_{ji}(t-1)} = -[R_{ref}(t-m) - R(t)]\frac{\partial R(t)}{\partial u(t-m)} \frac{\partial u(t-1)}{\partial T_{iN}}\frac{\partial T_{iN}}{\partial w_{ji}(t-1)} \tag{5.4.20}$$

$$\frac{\partial u(t-1)}{\partial T_{iN}} = (100 / P_B)[T / 2T_i^2(e(t-1) - e(t-2)) + (-T/T_i^2)e(t-1)]. \tag{5.4.21}$$

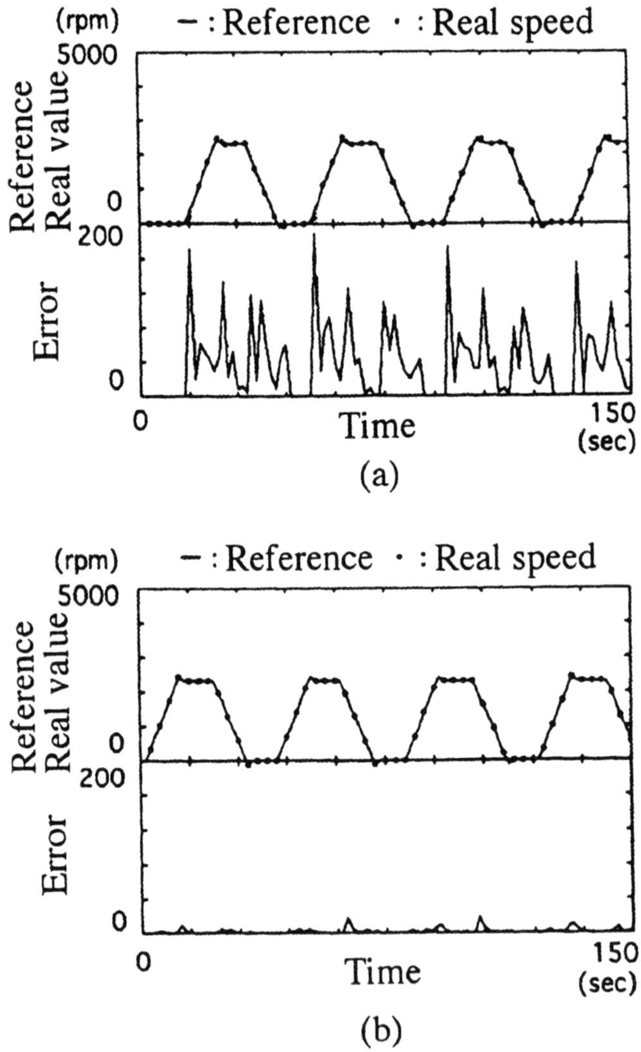

Fig. 5.4.4 Experimental results showing the speed control of the electric vehicle using: (a) conventional PID controller (b) self-tuning PID-type neuro-controller.

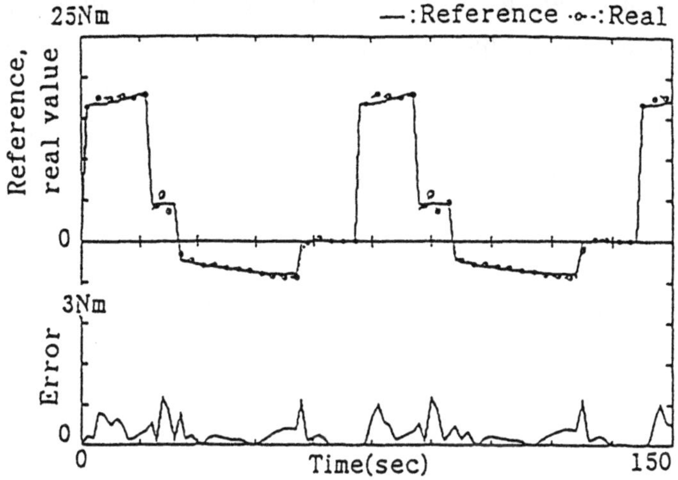

Fig. 5.4.5 Experimental result for motor torque control using the self-tuning PID-type neuro-controller.

5.4.5 Conclusions

Figures 5.4.4 (a) and (b) show the speed control performances of the system using a PID controller and the self-tuning PID-type neuro-controller, respectively. The PID controller was tuned using the tuning method of Takahashi et. al. Comparing the two results it can be observed that the neural network-tuned PID controller performed better with a reduced performance error. Figure 5.4.5. shows the motor torque control performance using the self-tuning PID-type neuro-controller. This experimental result can be compared to that of the conventional PID controller of Fig. 5.4.2. Similarly, it can be observed that the error has been reduced significantly using the proposed methodology.

This application has described a scheme where a neural network is used to tune a PID controller which is referred to as the self-tuning PID-type neuro-controller. The scheme has been applied for the speed control of an electric vehicle. Experimental results show that the actual error has been reduced considerably using the proposed scheme when compared to the conventional PID controller.

5.5 MIMO Furnace Control With Neural Networks

5.5.1. Introduction

In recent years, there have been a number of neuro-control learning schemes proposed in the literature [34]. Among these, the inverse model neuro-control approach, developed by Widrow and Stearns [16] and Psaltis *et al.* [17], has been one of the more viable techniques for implementation of neural networks in control. One reason for its utility is its simplicity. In the general learning approach of Psaltis *et al.* [17] as described in Section 4.2, a neural network is trained off-line similar to batch learning in pattern recognition where the target patterns are the plant inputs and the input patterns consist of the corresponding actual plant output and some delayed plant signals. In pattern recognition, the neural network is trained to memorize and recall the patterns, whereas, in control it is trained to learn the relationship of the patterns. Convergence is obtained by minimizing the error between the target pattern and the output of the neural network below a prescribed value. Once the network has learned the inverse model of the plant it is configured as a direct controller to the plant.

Although the inverse model neuro-control scheme is similar in architecture to a conventional feedback control scheme, the property of the neuro-controller is somewhat different from that of a conventional feedback controller. In effect, the neuro-controller has a property similar to that of a traditional self-tuning adaptive controller[1]. In the direct self-tuning control scheme, an *a priori* knowledge of the plant's mathematical model structure is required and the plant's parameters are needed to be estimated through an estimation scheme such as the popular least squares parameter estimation as described in Section 3.3. From the estimates of the plant's parameters, the regulator is able to generate the correct control input through an inverse mathematical model of the plant.

The inverse model neuro-control approach is more robust and simpler in this respect since it is not necessary to derive the mathematical model of the plant for implementation. The ability of neural network models to learn and generalize based on the input-output behavior of a process has a greater advantage where many control problems can now be easily solved with less precise advanced knowledge of the plant. The use of the semilinear sigmoid functions in the hidden layers of the neural networks has also made it possible for solving highly nonlinear control systems where many traditional adaptive and conventional control techniques have no solution. Moreover, it is applicable to complex and ill-defined plants when mathematical modeling is difficult.

Another advantage of using neural networks is that it can also be trained on-line to further improve its performance. In the basic inverse model approach, the neuro-controller can be trained on-line by minimizing the system's performance error. However, in this architecture, minimizing the performance error does not actually minimize the neuro-controller's output error because of the location of the plant. In order to minimize the actual output error of the neuro-controller, knowledge of the plant's Jacobian is needed, as proposed in the specialized learning scheme of Psaltis *et al.* [17], which would be contradictory to the advantage of the neuro-control approach as described earlier. In a study by Saerens and Soquet [35], the elements of the Jacobian of the plant may be approximated by their signs when the orientation in which the control parameters influence the outputs of the plant is known. This approach is acceptable when the plant to be controlled is simple and when the system has reached steady state.

A more accurate approach to train the neuro-controller on-line is to adapt the architecture of an indirect adaptive control scheme where the identification model is replaced by a neural network similar to the architecture as shown in Fig. 4.2.5. Narendra and Parthasarathy adopted this architecture where a reference model was also used [8]. A similar scheme, referred to as backpropagation-through-time has been developed by Werbos [36], Nguyen and Widrow [9], Jordan and Rumelhart [10], and also a few others. In this scheme a neural network model is first trained to emulate the plant. The neuro-emulator is then used as a means to backpropagate the performance error to obtain the equivalent error at the output of the neuro-controller. This scheme involves on-line realization of an initially untrained neuro-controller by applying random inputs and by continuously repeating the desired trajectory cycle until the performance error is minimized.

In this application, the neuro-emulator and neuro-controller architecture is adopted to control a MIMO miniaturized industrial furnace. Figure 5.5.1 illustrates the furnace and Fig. 5.5.2 shows its schematic diagram. In real-time it is impractical to realize both the neural network emulator and neuro-controller on-line as such a process is slowly time varying in nature. It is highly impractical to apply random inputs over a long period or to repeat a desired trajectory cycle over thousands of cycles. Thus, it is necessary to train the neural network models off-line initially.

Fig. 5.5.1. An illustration of the furnace.

The approach begins with judicious selection of neural network models which are trained off-line to learn the inverse and forward models of each channel of the furnace. The training patterns are obtained from the input-output behavioral characteristic of each channel of the process. These models are then tested on-line and the best neuro-controller and neuro-emulator models are chosen and configured to control the process in the architecture as shown in Fig.4.2.5. An advantageous feature of this configuration is that the neuro-controllers and neuro-emulators may be further trained in an on-line way to adapt to the environment and to improve their performance. However, in this research, there is no detailed analysis being done as regards to the behaviour of the neural networks.

One simple way to demonstrate the effectiveness of a newly-developed scheme is to make direct output comparisons with some existing control schemes. A self-tuning adaptive control scheme based on Clarke and Gawthrop algorithm [14], [15] and a conventional PID (proportional-plus-integral-plus-derivative) control schemes are [33] developed on the same furnace for comparison purposes. In the experiments, the three schemes are compared with respect to set-point changes, unknown load disturbances, parameter variations in the process, and saturation in the controller outputs. The experiments are also conducted to compare the performance of the neuro-control system without on-line and with on-line learning. A brief description of the furnace control system is given in the next section. In the following section, the implementation of neuro-control scheme on the furnace is explained. The experimental results are discussed and compared in the section that follows.

5.5.2 Description of Furnace Control System

The furnace control system is developed by Omron Inc., Japan, which consists of a three-channel copper cylinder (Fig. 5.5.1). The furnace is a miniaturized model for use in plastic moulding industry where the temperatures to be controlled in the three channels are between 100°C and 300°C. It consists of a sensor module, a parallel input-output (PIO) interface board, an NEC PC-9801F microcomputer, and an actuator. Figure 5.5.2 shows a schematic diagram of the furnace temperature control system. The heater capacities are 280 watts for the smaller cylinder which is called channel 1, and 1.2 kilowatts for the other two cylinders, channels 2 and 3. Although each of the channels can be controlled independently, the effect of heat flow from one channel to another makes the control rather difficult.

The sensor module includes a thermocouple of K type (TC12-4K) for measuring the output temperature of the cylinders. Temperatures in the range of 0 °C to 1,200 °C can be measured and converted to digital data by means of a 12-bit analog to digital converter (AD6940) with a resolution of 0.2 °C. The PIO interface board (PIO-16/16) is used as a means to transfer digital signals between the microcomputer and external devices and is inserted in the extended I/O slot of the microcomputer. Solid state thyristors

(Omron G3N-220B) are used to switch on and off the heaters according to the control signals. These signals are pulse width modulated (PWM). When the control signal is 100%, the heater is switched on continuously for 6 sec. in one sampling interval. The main control program is written using Microsoft-C, while the subroutine to operate the heaters is written in Assembly (Intel 8086).

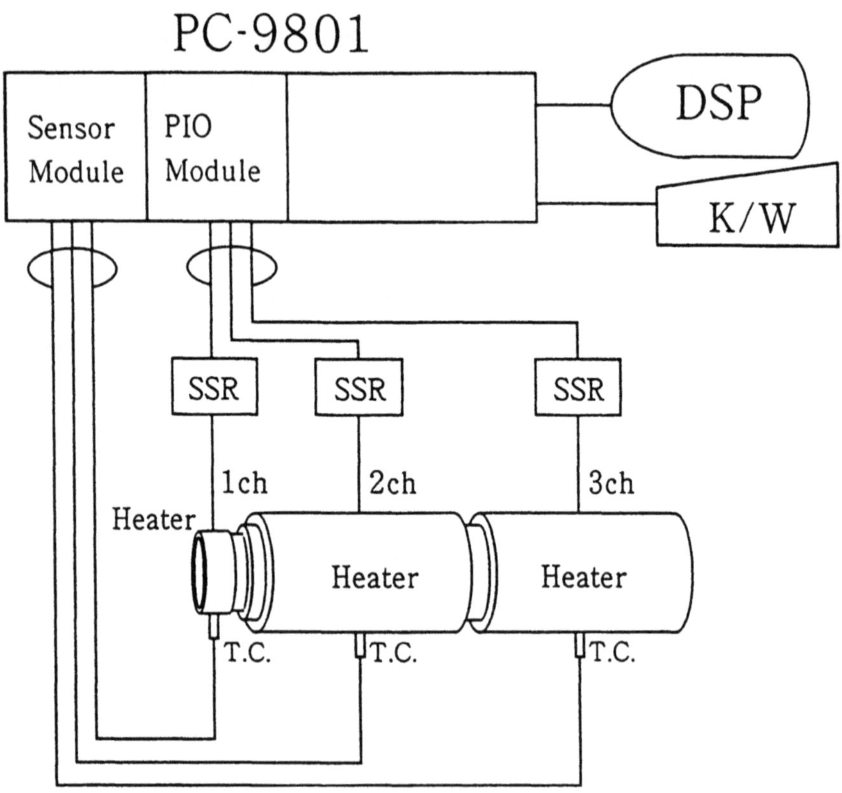

Fig. 5.5.2. Schematic diagram of the furnace control system.

5.5.3 The Neuro-Control Scheme

5.5.3.1 Off-line Training Scheme

Several neural network models having different numbers of input vector elements and hidden neurons are selected to develop the inverse and forward models of each channel of the MIMO furnace using the general learning architecture. As the neural networks are trained off-line, a selection of input-output training patterns is needed.

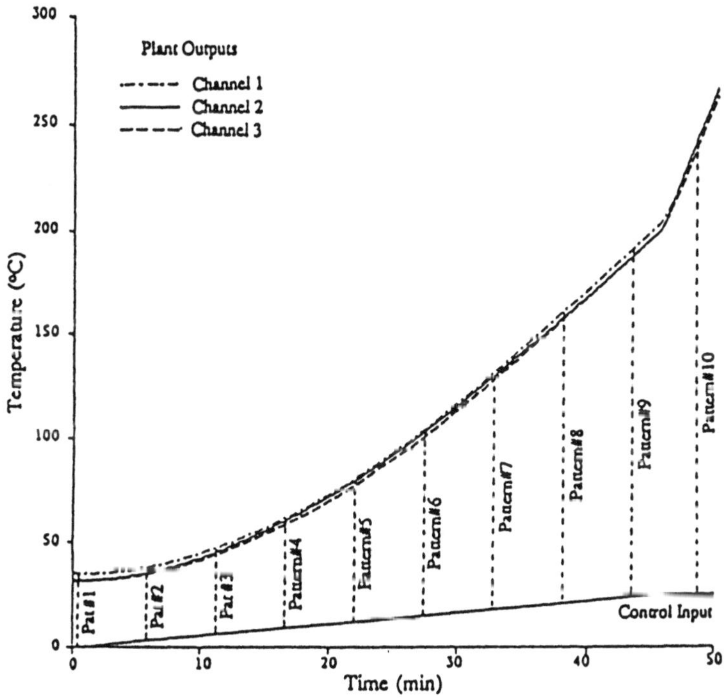

Fig.5.5.3. Open loop input-output characteristics of the furnace when a ramp input is injected. Ten pairs of input-output training patterns have been selected for training the neural network models.

In many developments (for example, [8]-[10]), the neural network models are trained to learn the inverse and forward models of the plants by injecting random input signals and iteration is done over thousands of cycles on-line until convergence is achieved. In our approach due to the slow nature of the process, which is, in addition open-loop stable, we injected ramp signals between the limits of the actuators. The open-loop input-output behavioral characteristics of each channel of the furnace can be observed as shown in Fig. 5.5.3.

Input-output training patterns for each channel are selected at regular intervals over the entire plant characteristic as shown in Fig. 5.5.3 to train the neural network models using the general learning scheme. The neuro-controller models are trained to learn the inverse dynamics of each channel of the furnace by using the respective channel's output data together with some corresponding delayed output values as the input patterns and the corresponding channel's input data as the target patterns. The neuro -emulator models are trained to learn the forward dynamics of each channel by using the input data together with some delayed output data as the input patterns and the corresponding output data as the target patterns. In training a neural network to learn an inverse or a forward dynamic model of a plant off-line, the backpropagation signal between the output and the hidden layers, is expressed as

$$\delta_k = (\tau_k - O_k) \tag{5.5.1}$$

where T_k is the target pattern and O_k is the actual output of the neural network, and between the hidden and input layers, it is expressed as

$$\delta_j = f'(S_j) \sum_k \delta_k w_{kj} \quad . \tag{5.5.2}$$

where f(x)= 1/(1+exp(-x)). The weights between the input and hidden layers are updated as

$$\Delta w_{ji}(t+1) = \eta \delta_j O_i + \alpha \Delta w_{ji}(t) + \beta \Delta w_{ji}(t-1) \tag{5.5.3}$$

and between the hidden and output layers as

$$\Delta w_{kj}(t+1) = \eta \delta_k O_j + \alpha \Delta w_{kj}(t) + \beta \Delta w_{kj}(t-1) \tag{5.5.4}$$

where O_i and O_j are the outputs of the input and hidden layers, respectively, η is the learning rate, and α and β are the momentum and acceleration coefficients, respectively. The constants η, α, and β are all chosen empirically.

The optimum number of training patterns, hidden neurons, and input vector elements of both the controller and neuro-emulator models are all chosen by experiments as there is still no reliable method of determining these parameters systematically or automatically. In order to avoid training a large number of neural network models we initially selected four models for each channel having 5, 10, 15 and 20 hidden neurons. The neural network models are then trained using several computers. The networks are then pruned around the best number of hidden neurons. The number of training patterns is also chosen by trial and error in a similar way. A large learning rate, η, is used initially at 0.01 and once oscillation occurs, it is reduced to a tenth of its previous value. The momentum terms, α and β, are chosen to be 0.8 and -0.15, respectively. For each model, training is stopped once convergence does not reduce by more than 0.0001 per cent over 1,000 iterations.

As the neural networks are highly nonlinear, it is difficult to determine analytically which model has learned the true plant inverse and forward dynamics. A simple and reliable method is to test each of the models on-line. For the controller, each neural network model is tested on-line by configuring it directly to control the process. The neuro-emulator models are tested by applying a ramp signal to each process and their respective emulators where their outputs are compared with each other.

We observed that different number of hidden neurons, input vector elements, and training patterns did have an effect on the neural network learning. The best neuro-controller for each channel consists of 10 hidden neurons and 3 input vector elements: the present plant output, y(t), and two delayed plant outputs y(t-1) and y(t-2). The best furnace's neuro-emulator of each channel consists of 12 hidden neurons with 3 input vector elements u(t), y(t-1), and y(t-2). It was found that a selection of ten training patterns as shown in Fig. 5.5.3 was suitable to obtain good convergence for training the neural network models.

5.5.3.2 On-line Learning

The respective neuro-controller and neuro-emulator of each process are configured in the control scheme as shown in Fig. 5.5.2. For this on-line scheme, the weights of the controller neural network are adapted using the performance error as follows

$$E = \frac{1}{2}[r - y]^2 \qquad (5.5.5)$$

where d and y are the desired and actual plant outputs, respectively. This error is backpropagated through the emulator and since its output neuron is a linear function, the error signal between the hidden and output layers of the emulator is

$$\delta_k^E = r - y \qquad (5.5.6)$$

where the superscript E denotes the emulator and the subscript k denotes the output layer. The error signal between the hidden and input layers of the emulator is

$$\delta_j^E = -\frac{\partial E}{\partial net_j^E} = \frac{\partial E}{\partial O_j^E} \cdot \frac{\partial O_j^E}{\partial net_j^E} \qquad (5.5.7)$$

where O_j^E and S_j^E are the output and input of the emulator's hidden layer neurons, respectively. By using the chain rule, the equation becomes

$$\delta_j^E = \left(\sum_j \partial_k^E \; w_{kj}^E \right) O_j^E \left(1 - O_j^E\right) \qquad (5.5.8)$$

where w_{kj}^E is the weight between the hidden and output layers of the emulator. The error signal between the hidden and output layers of the controller

$$\delta_k^C = \frac{\partial E}{\partial net_k^C} = \frac{\partial E}{\partial O_k^C} \frac{\partial O_k^C}{\partial net_k^C} \qquad (5.5.9)$$

where O_k^C and net_k^C are the output and input of the controller's output layer neurons, respectively. Since the output neuron of the controller is a saturating linear function

$$\frac{\partial O_k^C}{\partial net_k^C} = 1$$

thus,

$$\delta_k^C = \left(\sum_j \delta_j^E w_{ji}^E \right) \tag{5.5.10}$$

and for the hidden and input layers of the controller, the error signal is

$$\delta_j^C = \left(\sum_j \delta_k^C \, w_{kj}^C \right) O_j^C \left(1 - O_j^C\right) \tag{5.5.11}$$

where O_j^C is the output of the controller's hidden layer neurons. In this scheme, the weights of the neuro-emulator are not kept constant but are further improved on-line similar to the water bath scheme.

5.5.4. Experiments and Discussions

In many industrial control processes, temperature control is an important factor. The final product is affected if the temperature is too high or too low. Hence, the control target would be to bring the set-points during the startup as quickly as possible while avoiding large overshoots and maintaining it between the preset values. The experiments to control the furnace are performed with the purpose of achieving these objectives.

We conducted several experiments to test the robustness and adaptivity of the neural network based MIMO furnace control system. In order to see how well the neural furnace control system compares with some established control schemes, a multivariable self-tuning control and a conventional PID control schemes are also implemented on the same furnace. The multivariable self-tuning control scheme is based on the

Clarke and Gawthrop algorithm [14], [15] and the PID-control scheme is based on the velocity-form discrete-PID algorithm [33].

In implementing the self-tuning control scheme, an *a priori* mathematical model of the plant has to be derived. A more detailed derivation and development of the multivariable self-tuning furnace control system can be found in [36], [37]. Estimates of the model parameters of the furnace are obtained on-line by performing recursive least squares parameter estimation with a variable forgetting factor and a covariance matrix of diag {10, 10, 10}. A maximum period sequence which is a pseudo-random binary signal is used initially to speed up the convergence of the estimation of the parameters. The PID-controller is tuned using the method of Takahashi *et al.* [31] and minor adjustment is made to obtain the best response. Experiments are conducted to test the set-point tracking ability of the three control algorithms. Each experiment was conducted over 480 samples with a sampling interval of 10 seconds.

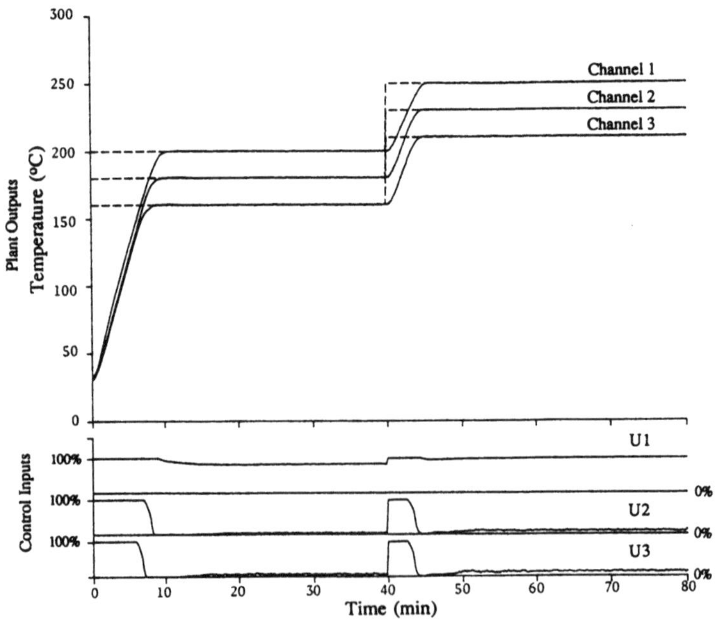

(a) neuro-control

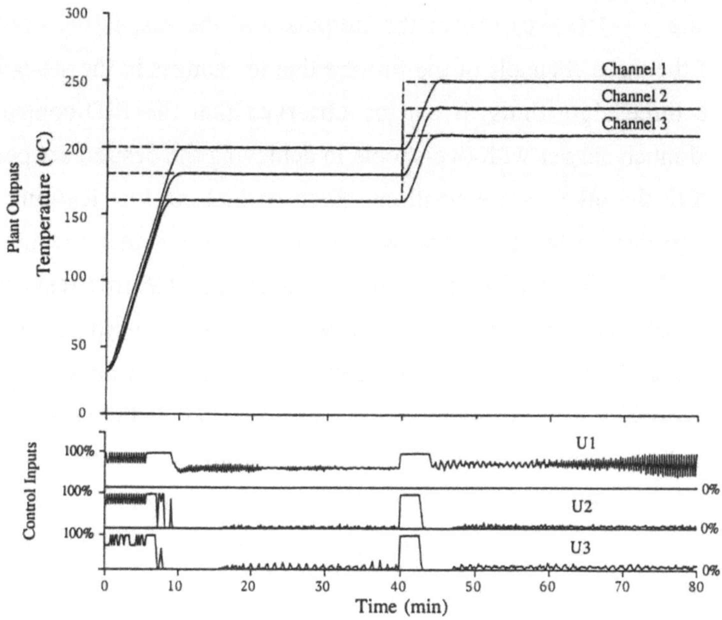

(b) self-tuning control

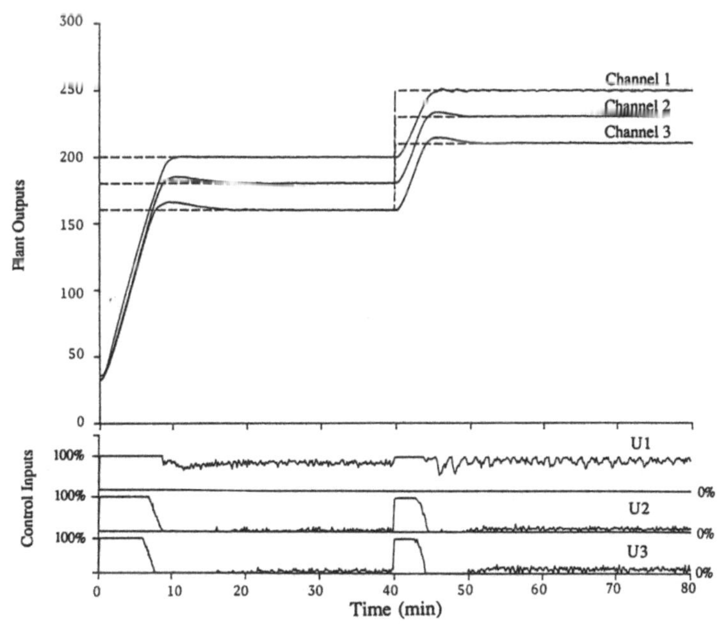

(c) PID control

Fig. 5.5.4. Tracking performance of the MIMO furnace using the three control schemes.

Figure 5.5.4 (a)-(c) shows the outputs and the respective control signals of the three channels of the furnace due to changes in the set-points using the three algorithms. It can be observed that the PID-controller performed much slower with overshoots in achieving the desired set-points compared to the other two algorithms. Even without on-line learning the neuro-controller performed quite well which indicates that the neuro-controller of each channel of the furnace has learned the true respective inverse model. With on-line learning the neuro-controller performed better where it was fine-tuned at every sample. The self-tuning controller performed equally well in following the desired set-points but selection of the user-defined polynomials was a rather time-consuming task. The user defined polynomials P and R are chosen as identity matrices whereas Q is diag $\{0, 0, 0.02\,(1-z^{-1})\}$. In this respect, once trained the neuro-controller is much easier to be implemented as there is no requirement for tuning or selection of user defined parameters.

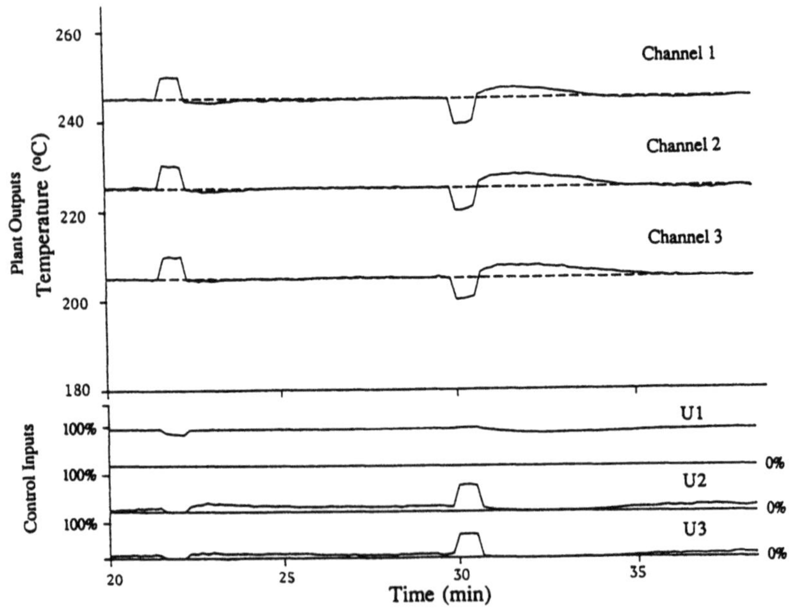

(a) neuro-control

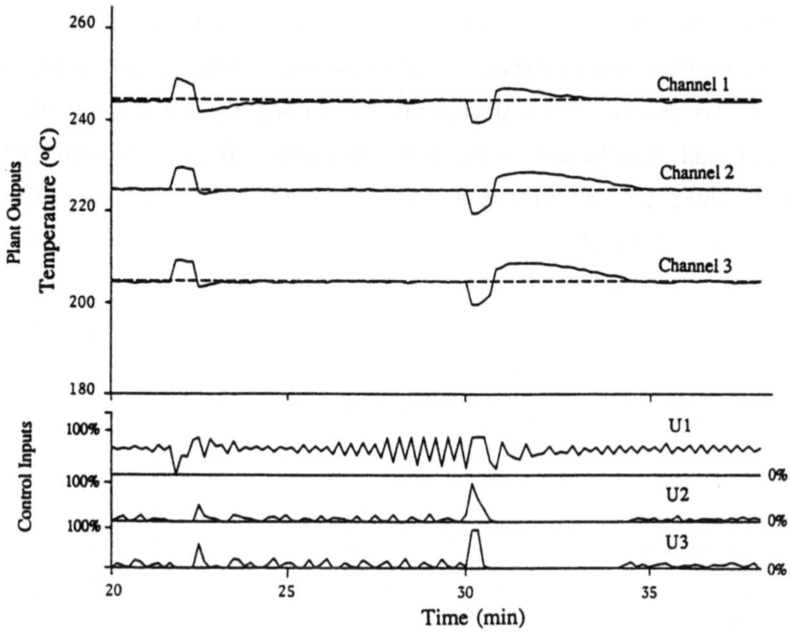

(b) self-tuning control

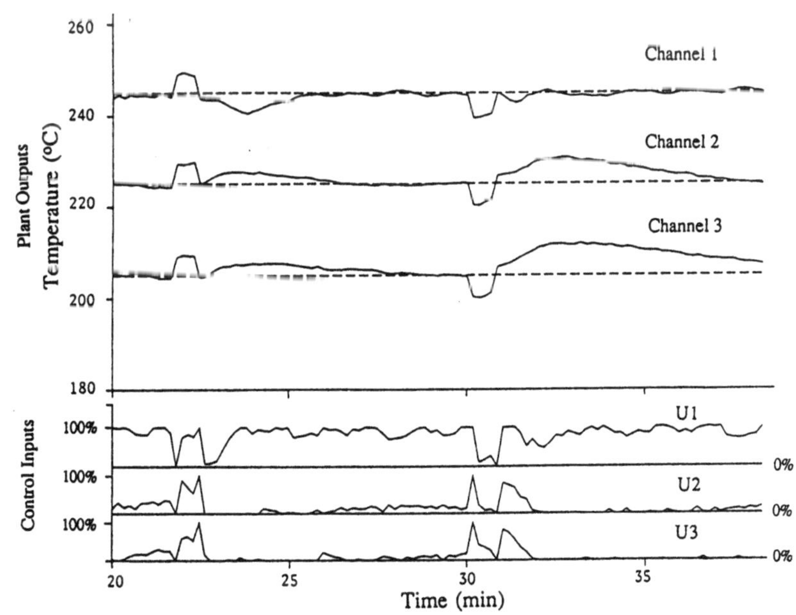

(c) PID control

Fig. 5.5.5. Performance of the MIMO furnace under the influence of load disturbances using the three control schemes.

We conducted a second set of experiments to study the ability of the three controllers when unknown load disturbances were imposed on the process. To ensure identical comparison among the three controllers, artificial load disturbances of value 5°C between $150 \le t \le 153$ and -5°C between $180 \le t \le 183$ were added to the outputs of each channel in the experiments. The performances of the three systems under the effects of these load disturbances are shown in Fig. 5.5.5 (a)-(c) at the sampling period of interest. It can be observed that the neuro-controller and the self-tuning controller performed equally well with the ability to recover quickly from the effects of the positive load disturbance. The neuro-controller was able to perform well due to its generalization capability and also its ability to adapt quickly to changes in its inputs. The self-tuning controller was also able to overcome this problem as it has a d-step ahead predictor. On the other hand, the PID-controller showed poor recovery which affected the system badly. In the case of the negative load disturbances the neuro-controllers and the self-tuning controllers were also affected as only positive control inputs are allowed, however, it can be observed that they recovered much more rapidly than the PID-controller.

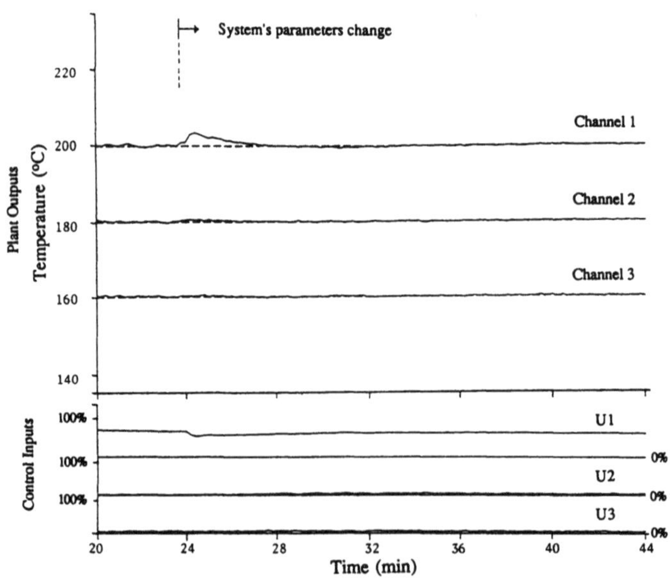

(a) neuro-control

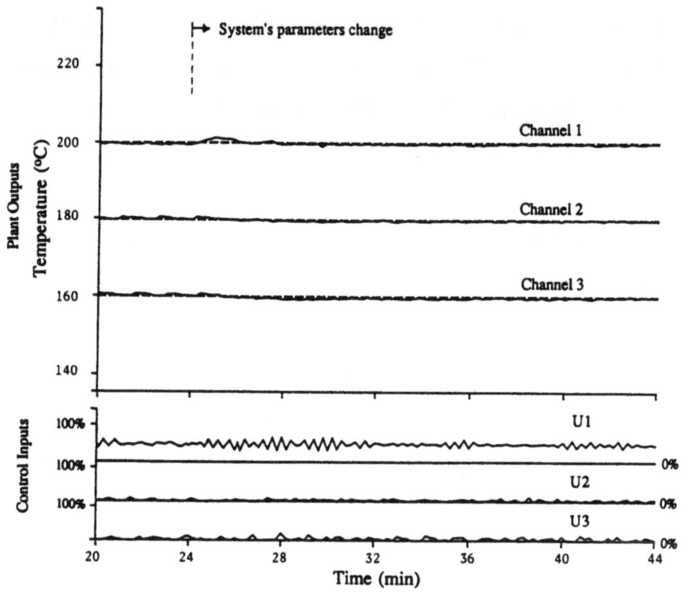

(b) self-tuning control

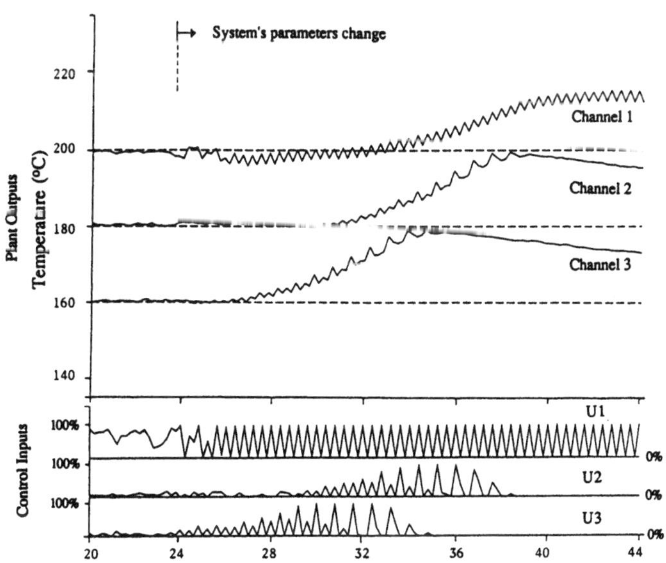

(c) PID control

Fig. 5.5.6. Behaviour of the MIMO furnace when there are parameter variations using the three control schemes.

The dynamic behaviour and parameters of many process control systems often tend to change in an unpredictable way during operation. To test the robustness of the three systems we conducted experiments on the furnace by multiplying a value of 0.03 to u(t-3) at the output of each channel after the 120th sampling instant, which represents a change in the process parameters. It is seen that these changes in dynamics are large and although it is difficult to imagine a real plant varying in such a drastic way, the experiments were conducted to illustrate the relative robustness and adaptivity of the controllers.

The behaviour of the three controllers on the furnace under this influence is shown in Fig. 5.5.6 (a)-(c). It can be observed that the PID controller was badly affected when the parameters of the system changed. Despite the integral action, offset is seen in all channels due to persistent control saturation.

The self-tuning ability of the multivariable self-tuning controller proves to be a distinct advantage over the conventional PID-controller in that it can adapt to changes in the plant parameters. Similarly, the neuro-controller also showed considerable robustness in this respect due to its powerful generalization capability coupled with the adaptive on-line learning.

We conducted further tests on the three controllers by imposing strong saturation constraints in their outputs. Nonlinear saturation where the control signal dropped to zero below 20% and clipped at 80% was imposed on each controller. The same experiments were conducted as in the third set of experiments where a change in the process parameters was also experienced. The behavior of the neuro-controller without on-line learning is shown as in Fig. 5.5.7 (a) where steady-state errors can be observed at the outputs of channels 2 and 3. When there was a change in the plant parameters the performance of the neuro-controller worsens. Figure 5.5.7 (b) shows how the neuro-controller behaved when there was on-line learning. It can be observed that even after the plant parameters have changed the neuro-controller still performed quite well.

Figure 5.5.7 (c) shows the behavior of the self tuning controller under the same experimental conditions. The self-tuning controller performed

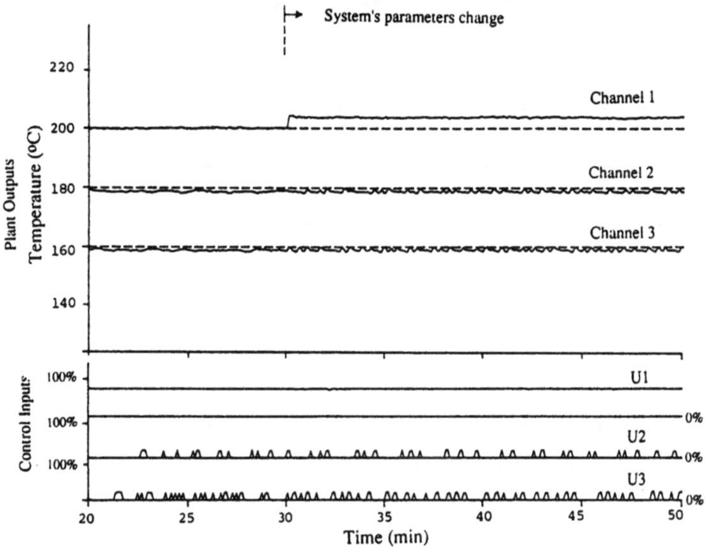

(a) neuro-control without on-line learning

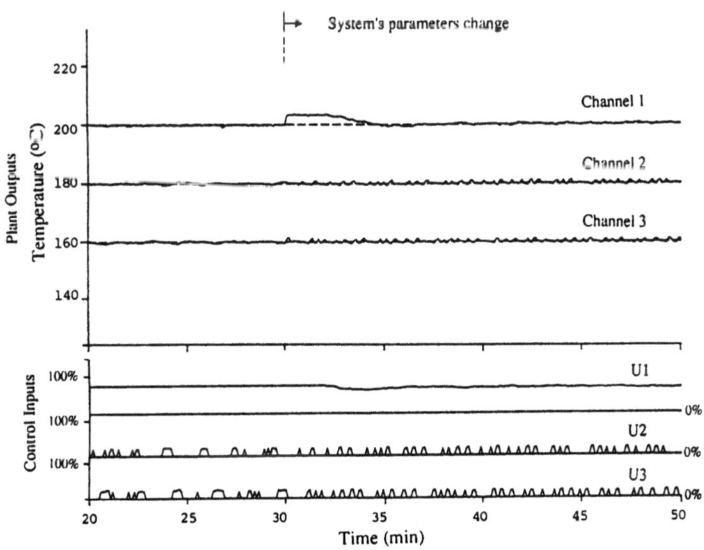

(b) neuro-control with on-line learning

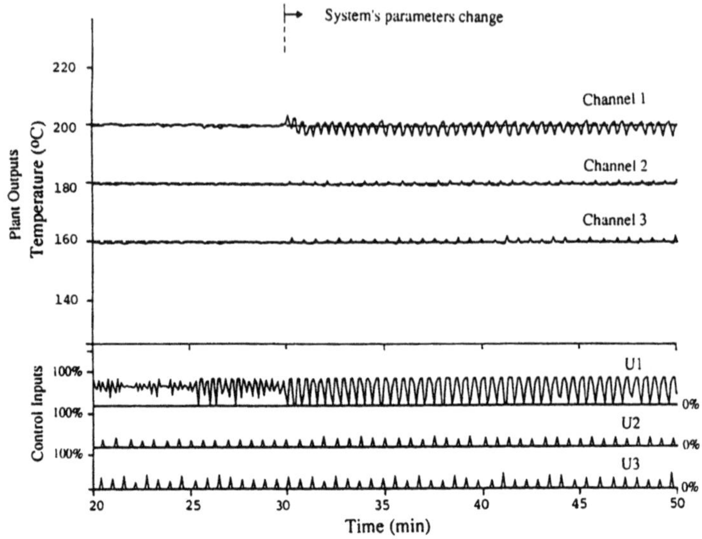

(c) self-tuning control

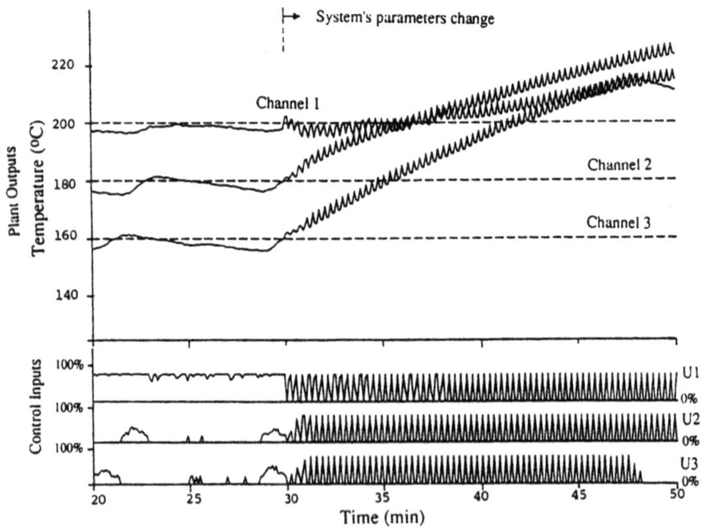

(d) PID control

Fig. 5.5.7. Behaviour of the MIMO furnace when there are parameter variations and the control signal are subjected to strong nonlinear saturations using four different control schemes.

well before there was a change in the plant parameters. However, when the plant parameters changed the performance of the controller deteriorated, especially that of channel 1. In the case of the PID-controller, it can be observed that this conventional controller cannot cope with such strong saturation constraints in the control signals. The process became unstable as shown in Fig. 5.5.7(d) when the parameters of the plant changed and when the control signals are driven into saturation.

From all the experiments, it can also be observed that the outputs of the neuro-controller were very smooth in nature even during the occurrence of large load disturbances. On the other hand, the outputs of the self-tuning and the PID controllers fluctuated adversely which could shorten the lifetime of the actuators under long period of usage.

The control signals of the self-tuning controller are rather oscillatory in nature, especially, that of channel 1 as the channel is noisier and has a shorter time-delay. The objective of the self-tuning control approach is minimizing the variance of the plant output, and as the plant is rather noisy the variance is increased which results in oscillatory control signals. This also leads to degradation in performance of the self-tuning controller when the control signal is subjected to saturation constraints as in the last set of experiments. The objective of the neuro-controller on the other hand is learning the pseudo-inverse of the plant. The nonlinear mapping ability of the semilinear neurons in the hidden layers of the neural networks leads to smoother controller outputs which could prolong actuator life.

5.5.5 Conclusions

A neuro-control scheme consisting of neuro-emulators and neuro-controllers has been implemented in real-time to control a MIMO furnace. The approach undertaken can be easily applied to a wide variety of open-loop stable processes without the use of any initial conventional controller or knowledge regarding dynamics. The on-line learning scheme allows both the neuro-emulators and neuro-controllers to adapt to changes in the environment and to further improve their performances.

The neuro-control scheme developed on the furnace has also been compared to a self-tuning control and a conventional digital-PID control schemes through several experiments. The experimental results show that the neuro-controller is superior and has considerable robustness. In comparison to effort of implementation among the three schemes, the digital-PID-controller requires the least effort. However, the experimental results show that this conventional controller cannot cope with complexities in the plants.

Between the self-tuning control and the neuro-control schemes, the latter requires relatively less effort. Although in the neuro-control scheme the forward plant model is required to backpropagate the performance error to get the equivalent error at the controller's output, the development of the neuro-emulator for the plant is unduly expensive. The neuro-emulator learns the plant model from input-output observations of the plant, whereas, in the self-tuning control scheme, a mathematical model is needed to be derived. Even if the neural network does not learn the forward model well, the emulator can still be used as a means to backpropagate the performance error, however, at the expense of longer on-line learning cycles. In the case of the self-tuning controller, inaccurate modeling of the plant can lead to degraded performance, especially when wrong time-delay is observed.

One drawback in the neuro-control scheme is the requirement for prior training, and that a judicious selection of neural network models is essential for its success. However, it is inexpensive to train several neural network models at the same time using several computers and pruned around the model that exhibits the best performance. Moreover, once trained, the neuro-controllers have no requirement for tuning as required for conventional PID controllers. In the self-tuning control scheme, and also in many other adaptive control schemes, there is a requirement for prior selection of user-defined parameters which is a difficult and time consuming task, even though the controller is self-tuned.

This study has shown that neural network may become an important technology for many industrial control applications. The simplicity and reliability of neuro-control over traditional control techniques will be the key for the development of efficient and intelligent control systems in the near future.

5.6 Concluding Remarks

This chapter describes four kinds of neuro-control application. The first application involves implementation of neuro-control on a water bath process. The robustness of this neuro-control water bath temperature control system is tested under varying complexities such as parameter variation, variable time-delay, and load disturbances in the process. The neuro-control scheme is compared to three other traditional control schemes; namely, fuzzy logic control, generalized predictive control, and PID control. The neuro-control scheme showed considerable robustness and has encouraging advantages over the other schemes.

The comparative study of the four control schemes on the first application is verified on another application, a MIMO furnace. In this application the same neuro-control strategy has been applied and which uses a neuro-controller and a neuro-emulator. This neuro-control scheme has been compared to a self-tuning control and PID control schemes implemented on the same furnace. The experimental results show that neuro-control furnace system has more robustness than the other two schemes especially in reducing noise and overcoming non linearities.

An interesting and advantageous feature of neural networks is their ability to learn and generalize. The same neuro-control strategy has been adopted in developing the water bath and also the furnace neuro-control systems. A novel feature of the neural network approach is that the same back-error-propagation algorithm can be used to solve two different control problems using only the plants' open loop input-output characteristics. In many traditional adaptive control techniques, however, it is a prior requirement to derive individually the mathematical models of the plants as accurately as possible before they can be implemented.

In the inverted pendulum system, a neural network is used to improve the limitations of the linear optimal controller. Obviously the linear optimal controller could not overcome nonlinearities such as friction and backlash in the system. The neuro-controller has been used as a compensator to the inverted pendulum system.

In another application, a neural network has been used to tune the parameters of a PID controller which we have referred to as the self-tuning PID neuro-controller. Such application of neural networks have tremendous potential in many existing traditional control approaches. For example, many adaptive controllers have a number of parameters or user defined functions that are needed to be selected in prior before they can be implemented. Especially when the plant is complex, the task of selecting these parameters, which is usually done by trial and error, becomes difficult and laborious. By integrating the neural network learning capability into the existing control startegy, such laborious procedure can be overcome.

5.7 References

[1] Åström, K.J. and B. Wittenmark, *"Adaptive Control"*, Addison Wesley, USA, 1989.

[2] Narendra, K.S., R. Ortega, and P. Dorato, Eds., *"Advances in Adaptive Control"*, IEEE Press, New York, 1991.

[3] "Special Issue on Neural Network Hardware", *IEEE Trans on Neural Networks* , Vol. 3, pp. 347-506, 1992.

[4] Togai, M. and H. Watanabe, "Expert system on a chip: An engine for real-time Approximate Reasoning", *IEEE Expert*, Vol. 1, pp. 55-62, 1986.

[5] Yamakawa, T., "Fuzzy microprocessors - rule chip and defuzzification chip", *Proc. of Int. Workshop on Fuzzy Systems Applications*, Iizuka-88, Kyushu, pp. 51-52, 1988.

[6] Kong, S.G. and B. Kosko, "Adaptive fuzzy systems for backing up a truck-and-trailer", *IEEE Trans. on Neural Networks* , Vol. 3, pp. 211-223, 1992.

[7] Kraft, L.G., and D.P. Campagna, "A comparison between CMAC neural network control and two traditional adaptive control systems", *IEEE Control Systems Magazine,* Vol. 10, pp. 36-43, 1990.

[8] Narendra, K.S. and K. Parthasarathy, "Identification and control of dynamical systems using neural networks", *IEEE Trans. on Neural Networks,* Vol. 1, pp. 4-27, 1990.

[9] Nguyen, D.H., and B. Widrow, "Neural networks for self-learning control systems", *IEEE Control Systems Magazine,* Vol. 10, pp. 18-23, 1990.

[10] Jordan, M.I. and D.E. Rumelhart, "Forward models: supervised learning with a distal teacher", Cognitive Science, Vol. 16, pp. 313-355, 1992.

[11] Moore, C.G. and C.J. Harris, "Indirect adaptive fuzzy control", *Int. Journal of Control.*, Vol. 56, pp. 441-468, 1992.

[12] Procyk, T.J. and E.H. Mamdani, "A Linguistic self-organizing process controller", *Automatica*, Vol. 15, pp. 15-30, 1979.

[13] Berenji, H.R. and P. Khedkar, "Learning and tuning fuzzy logic controllers through reinforcements", *IEEE Trans. on Neural Networks*, Vol. 3, pp. 724-740, 1992.

[14] Clarke, D.W., C. Mohtadi, and P.S. Tuffs,"Generalized predictive control-part I: The basic algorithm", *Automatica*, Vol. 23, pp.137-148, 1987.

[15] Clarke, D.W., C. Mohtadi, and P. S. Tuffs, "Generalised predictive control-part II: Extensions and interpretations", *Automatica*, Vol. 23, pp.149-160, 1987.

[16] Widrow, B. and S.D. Stearns, *"Adaptive Signal Processing"*, Prentice Hall, Englewood Cliffs, New Jersey, 1985.

[17] Psaltis, D., A. Sideris, and A. Yamamura,"A multilayered neural network controller", *IEEE Control Systems Magazine,* Vol. 10, pp.44-48, 1988.

[18] Rumelhart, D.E., G.E. Hinton, and R.J. Williams, "Learning internal representations by error propagation", in *Parallel Distributed Processing: Explorations in the Microstructure of Cognition Vol.I,* MIT Press, Cambridge, MA, pp. 318-362, 1986.

[19] Khalid, M. and S. Omatu, "A neural network controller for a temperature control system", *IEEE Control Systems Magazine*, Vol. 12, pp.58-64, 1992.

[20] Yusof, R. and S. Omatu, "Application of generalised predictive control to a temperature control process", *Proc. of IECON '91*, Kobe, Japan, Vol. 3, pp. 1935-1940, 1991.

[21] Kuperstein, M., "Neural model of adaptive hand-eye co-ordination for single postures", *Science*, Vol. 239, pp. 1308-1311, 1988.

[22] Nagata, S., M. Sekiguchi, and K. Asakawa,"Mobile robot control by a structured hierarchical neural network", *IEEE Control Systems Magazine,* Vol. 10, pp. 69-76, 1990.

[23] Zadeh, L.A., "Fuzzy sets", *Information and Control,* Vol. 8, pp. 338-352, 1965.

[24] Zadeh, L.A., "Outline of a new approach to the analysis of complex systems and decision processes", *IEEE Trans. on Systems, Man, and Cybernetics*, Vol. 3, pp.28-44, 1973.

[25] Tong, R.M., "A control engineering review of fuzzy systems", *Automatica*, Vol. 13, pp. 559-569, 1977.

[26] King P.J., and E.H. Mamdani, "The application of fuzzy control systems to industrial processes", *Automatica*, Vol. 13, pp. 235-242, 1977.

[27] Assilian, S. and E.H. Mamdani, "An experiment in linguistic synthesis with a fuzzy logic controller", *Int. Journal of Man-Machine Studies*, Vol. 7, pp.1-13, 1974.

[28] Kosko, B., *"Neural Networks and Fuzzy Systems: A Dynamical Systems Approach to Machine Intelligence"*, Prentice-Hall, Englewood Cliffs, New Jersey, 1991.

[29] Linkens, D.A. and M.F. Abbod, "Self-organizing fuzzy logic control and the selection of its scaling factors", *Trans. on Instrumentation, Measurement and Control*, Vol. 14, pp. 114-125, 1992.

[30] Daley, S. and K.F. Gill, "A design study of a self-organising fuzzy logic controller", *Proc. of I. Mech E*, Vol. 200, (CI), pp.59-69, 1986.

[31] Takahashi, Y., C.S. Chan, and D.M. Auslander,"Parametereinstellung bei linearen DDC-algorithmen", *Regelungstechnik und Prozeß- Datenverarbeitung*, Vol. 19, pp.237-244, 1971.

[32] Levin, E., R. Gewirtzman,and G.F. Inbar, "Neural network architecture for adaptive system modelling and control", *Proc. of Int. Joint Conference on Neural Networks (IJCNN) '89*, Washington D.C., Vol. II, pp. 311-316, 1989.

[33] Ogata, K.,*"Modern Control Systems"*, Prentice-Hall, Englewood Cliffs, New Jersey, 1989.

[34] Miller III, W.T., R.S. Sutton, and P.J. Werbos, Eds., *"Neural Networks for Control"*, MIT Press, Cambridge, MA, 1990.

[35] Saerens, M. and A. Soquet, "A neural controller based on backpropagation algorithm", *Proc. of First IEE Int. Conf. on Artificial Neural Networks,* London, pp.211-215, 1989.

[36] Yusof, R., S. Omatu, and M. Khalid, "Self-tuning PID control: A multivariable derivation and application", *Automatica* (to be published).

[37] Yusof, R., "Theory and applications of self-tuning PID control and generalized predictive control", Ph.D. Thesis, University of Tokushima, Japan, 1994.

PROGRAM LIST

```
/* This program can be used to train a neural network model with */
/* one hidden layer to learn the inverse dynamics of the plant */
/* Once trained the neural network model can  */
/* be used as a direct controller to the plant */
/* The plant input-output characteristics must first be  */
/* generated and kept in a file called in-out.dat */
/* This program runs under the Borland C++ Version 3.1 Environment */


/* Backpropagation Neuro-Control*/
#include <stdio.h>
#include <math.h>
#include <string.h>
#include <stdlib.h>

/* Dcclare the total number of nodes and pattern */
#define No_of_Input_Units     3  /* No. of nodes of input layer i */
#define No_of_Hidden_Units    15 /* No. of nodes of hidden layer j */
#define No_of_Output_Units    1  /* No. of node s of output layer k */
#define No_of_Patterns        8  /* Total no. of patterns p */

#define Eta            0.0001 /* The learning rate */
#define Alpha          0.8    /* Momentum constant */
#define Beta           -0.15  /* Acceleration constant */
#define ErrorFunc      0.001  /* Maximum error allowed */
#define Wmin           -0.05  /* For use in random function */
#define Wmax           0.05   /* to generate random values */
 #define No_of_Data    101    /*Total number of plant input-output data */

/* Functions */
#define f(x)       (1/(1+exp(-(x))))
                   /* Sigmoid function where x = net + bias*/
#define rnd( )     ((float)rand( )/0x7fff * (Wmax-Wmin) + Wmin)
                   /* This function is used to generate random values */

/* Declare outputs, weights, and biases */
float     o1[No_of_Patterns][No_of_Input_Units];
          /* Output of Input layer i */
float     o2[No_of_Hidden_Units];     /* Output of Hidden laye r j */
float     o3[No_of_Output_Units];     /* Output of Output layer k */
float     t[No_of_Patterns][No_of_Output_Units];
```

```
            /* Desired Output or target value */
float       w21[No_of_Hidden_Units][No_of_Input_Units];
            /* Weights from i to j layer */
float       dw21[No_of_Hidden_Units][No_of_Input_Units][3];
            /* Change in the above weights */
float       w32[No_of_Output_Units][No_of_Hidden_Units];
            /* Weights from j to k layer */
float       dw32[No_of_Output_Units][No_of_Hidden_Units][3];
            /* Change in the above weights */
float       bias2[No_of_Hidden_Units];
            /* Bias to nodes in layer j (theta j) */
float       dbias2[No_of_Hidden_Units][3];
            /* Change in the above bias j */
float       bias3[No_of_Output_Units];
            /* Bias to nodes in layer k (theta k) */
float       dbias3[No_of_Output_Units][3];
            /* Change in the above bias j */
float       increase, erorfunc, er[3], u[No_of_Data], y[No_of_Data];
float       Errorfunction
int         count, iterations;


/* Main Program */
void  main ()
/*begin main program*/
{
  unsigned long           i,j,k;
  static float        errorfunc;
  void                propagation(int);
  void                back_propagation(int);
  void                forward(int);
  void                read_input-output_data( );
  void                initialize( );
  void                save_weight( );
  void                NN_inputs( );

  initialize( );
  read_input-output_data( );
  increase=0.0;
  for (errorfunc = 0.0, i=0; i<No_of_Patterns;i++)
     {
          forward(i);
          for(j=0; j<No_of_Output_Units; j++)
             {
          errorfunc += pow(t[i][j] - o3[j], 2.0);
                 printf("Error Function is: %.5f\n", errorfunc);
                 /* error is the square of target - output */
             }
     }
  errorfunc /= 2;
          erorfunc=errorfunc;
```

```
        printf("Error Value is : %.3f\n", errorfunc);
        /* print the error value */
        er[0]=errorfunc;

 for( i=0; errorfunc>ErrorFunc;)
    {
        for( j=0; j<No_of_Patterns; j++ )
        {
           propagation(j);
           back_propagation(j);
        }
        for( errorfunc=0.0, j=0; j<No_of_Patterns;j++)
           {
                forward(j);
                for(k=0; k<No_of_Output_Units; k++)
                   errorfunc += pow(t[j][k] - o3[k], 2.0);
           }

        iterations=1; ++i;
        increase+=(float)iterations;
        errorfunc /= 2;  erorfunc=errorfunc;
        {
           printf(" I= %f ",increase);
           printf("E = %.7f\n", errorfunc);
                er[1]=er[0];  er[0]=errorfunc;
           if(er[1]<er[0])break;
        }
    }

        printf("\n  After Learning   \n");
        for( i=0; i<No_of_Patterns; i++)
        forward(i);
        save_weight();
 };



/* To save the adapted weights in a data file */
void       save_weight()
{
  int      i,j;
  FILE *fp;
 fp = fopen("weight.dat","w");
         /* To save the weights in layer j */
         for( i=0; i<No_of_Hidden_Units; i++)
            for( j=0; j<No_of_Input_Units; j++)
         {
                 fprintf(fp,"%f\n",w21[i][j]);
                 fprintf(fp,"%f\n",bias2[i]);
         }

         /* To save the weights in layer k */
         for( i=0; i<No_of_Output_Units; i++)
```

```
        for( j=0; j<No_of_Hidden_Units; j++)
          {
                fprintf(fp,"%f\n",w32[i][j]);
                fprintf(fp,"%f\n",bias3[i]);
          }
        fclose(fp);
}


/* Procedure Propagation */
/* To propagate the network forward */
void    propagation(int p)
{
        int     i,j;
        float   net;

  for( i=0; i<No_of_Hidden_Units; /* layer j */ i++)
     {  for( net=0.0, j=0; j<No_of_Input_Units; j++)
                net += w21[i][j] * o1[p][j];
                o2[i] = f(net+bias2[i]);        /* sigmoid function */
     }

  for( i=0; i<No_of_Output_Units; /* layer k */ i++)
     {
        for( net=0.0, j=0; j<No_of_Hidden_Units; j++)
             net += w32[i][j] * o2[j];
             o3[i] = (net+bias3[i]);
             /* output of NN not bounded by sigmoid function */
     }
  }



/* Procedure Back Propagation */
/* To backpropagate error through the network */
void    back_propagation (int p)
{
        int     i,j;
        float   d2[No_of_Hidden_Units];     /* d2 is delta j */
        float   d3[No_of_Output_Units];     /* d3 is delta k */
        float   sum;            /* sum is add. of delta k * Wkj */

  /* to calculate delta k */
  for( i=0; i < No_of_Output_Units; i++ )
        d3[i] = (t[p][i] - o3[i]);

  for( i=0; i<No_of_Hidden_Units; i++)
     {

        /* to calculate the weights, w and dw, between layer j and k */
        for(sum=0.0, j=0; j<No_of_Output_Units; j++)
           {
```

```
                dw32[j][i][0] = Eta * d3[j] * o2[i]
                        + Alpha * dw32[j][i][1] + Beta * dw32[j][i][2];
                w32[j][i] += dw32[j][i][0];
                sum += d3[j] * w32[j][i];
                dw32[j][i][2] = dw32[j][i][1];
                dw32[j][i][1] = dw32[j][i][0];
            }

        /* to calculate delta j */
        d2[i] = o2[i] * (1 - o2[i]) * sum;
    }

  /* to calculate bias k */
        for( i=0; i<No_of_Output_Units; i++)
        {
                dbias3[i][0] = Eta * d3[i] + Alpha * dbias3[i][1]
                        + Beta * dbias3[i][2];
                bias3[i] += dbias3[i][0];
                dbias3[i][2] = dbias3[i][1];
                dbias3[i][1] = dbias3[i][0];
        }

  /* to calculate the weights, w and dw, between layer j and i */
  for( i=0; i<No_of_Input_Units; i++)
        for( j=0; j<No_of_Hidden_Units; j++ )
        {
                dw21[j][i][0] = Eta * d2[j] * o1[p][i]+ Alpha
                        * dw21[j][i][1] + Beta * dw21[j][i][2];
                w21[j][i] += dw21[j][i][0];
                dw21[j][i][2] = dw21[j][i][1];
                dw21[j][i][1] = dw21[j][i][0];
        }

  /* to calculate bias j */
  for ( i=0; i<No_of_Hidden_Units; i++ )
        {
                dbias2[i][0] = Eta * d2[i] + Alpha * dbias2[i][1]
                        + Beta * dbias2[i][2];
                bias2[i] += dbias2[i][0];
                dbias2[i][2] = dbias2[i][1];
                dbias2[i][1] = dbias2[i][0];
        }
}


/* Procedure to read input-output characteristics of systems */
void read_input-output_data( )
{
  int     j,k;
  FILE *file;
        file = fopen("in-out.dat","r");
```

```
/* this file must be generated which consists
   of plant input-output characteristics */

  for(j=0; j < No_of_Data; j++)
      fscanf(file,"%f %f\n",&u[j],&y[j]);

  for(j=0; j < No_of_Data; j++)
      printf("u[%d]=%f y[%d]=%f\n",j,u[j],j,y[j]);
  fclose(file);
}



/* Procedure to equate NN inputs */
void NN_inputs()
 /* to select the patterns over entire
    input-output plant characteristics */

{
         int      i, j, k;

   for(i=0,j=0; j < No_of_Patterns; j++)
      {
          for(k=0; k <No_of_Output_Units; k++)
                 t[j][k]=u[i+2];
      }

  for(j=0; j < No_of_Patterns; j++)
      for(k=0; k <No_of_Output_Units; k++)
           printf("T[%d][%d]=%f\n", j, k, t[j][k]);

   for(i=0, j=0; j < No_of_Patterns; j++)
 /*example for 3 NN input vectors */
      {    o1[j][0]=y[i+2];
           o1[j][1]=y[i+1];
           o1[j][2]=y[i];
      }

    for(j=0; j < No_of_Patterns; j++)
        for(k=0; k <No_of_Input_Units; k++)
            printf("o1[%d][%d]=%f\n",j,k,o1[j][k]);
  }



  /* Procedure forward*/
    void   forward(int p)
  {
           int     i;

           printf("\t%d ->", p+1);
           propagation(p);
           for( i=0; i<No_of_Output_Units; i++ )
```

```
            printf ("  o3[%d]=%2.5f\n",p+1, o3[i]);
}



 /*  Procedure Initialize */
/*  To randomly initialize the weights */
void      initialize( )
{
          int      i,j;

  /* To initialize the weights between layer j and i */
          for ( i=0; i<No_of_Hidden_Units; i++)
          {         for( j=0; j<No_of_Input_Units; j++)
                              w21[i][j] = rnd();
                              bias2[i] = 1.0;
          }

  /* To initialize the weights between layer k and j */
          for ( i=0; i<No_of_Output_Units; i++)
          {         for( j=0; j<No_of_Hidden_Units; j++)
                    w32[i][j] = rnd();
          bias3[i] = 1.0;
          }
}
```

INDEX

MIX
Papier aus verantwortungsvollen Quellen
Paper from responsible sources
FSC® C105338

If you have any concerns about our products,
you can contact us on
ProductSafety@springernature.com

In case Publisher is established outside the EU,
the EU authorized representative is:
Springer Nature Customer Service Center GmbH
Europaplatz 3, 69115 Heidelberg, Germany

Printed by Libri Plureos GmbH
in Hamburg, Germany